# Protecting Biolog Diversity

## Roles and Responsibilities

Edited by

CATHERINE POTVIN,
MARGARET KRAENZEL,
AND GILLES SEUTIN

McGill-Queen's University Press
Montreal & Kingston · London · Ithaca

© McGill-Queen's University Press 2001
ISBN 0-7735-2158-5 (cloth)
ISBN 0-7735-2159-3 (paper)

Legal deposit second quarter 2001
Bibliothèque nationale du Québec

Printed in Canada on acid-free paper

Publication of this book has been made possible through
a grant from the International Development Research
Centre, Biodiversity Division.

McGill-Queen's University Press acknowledges the
financial support of the Government of Canada through
the Book Publishing Industry Development Program
(BPIDP) for its activities. It also acknowledges the
support of the Canada Council for the Arts for its
publishing program.

**National Library of Canada Cataloguing in Publication Data**

Main entry under title:

Protecting biological diversity: roles and responsibilities

Papers originally presented at a symposium held during
the IUCN World Conservation Congress, Montreal, Quebec,
13–23 Oct. 1996.
Includes bibliographical references.
ISBN 0-7735-2158-5 (bound) – ISBN 0-7735-2159-3 (pbk.)

1. Biological diversity conservation—Developing countries—
Congresses. 2. Biological diversity conservation—Moral and
ethical aspect—Developing Countries—Congresses. I. Potvin,
Catherine, 1957– II. Kraenzel, Margaret III. Seutin, Gilles, 1961–
IV. World Conservation Congress (1st : 1996 : Montreal, Quebec)

QH75.P773 2001    333.95'16'091724    C2001-900202-5

Typeset in 10.5/13 Sabon by True to Type

## Protecting Biological Diversity
### Roles and Responsibilities

The most species-rich regions of the globe, the tropics, are economically the poorest. How can biologists work toward effective protection for endangered species in countries hungry for food and basic resources? And why should local people in those countries trust the advice of scientists from wealthier countries, who have broken their promises in the past and have typically shown little respect for the cultural values of others?

Catherine Potvin, Margaret Kraenzel, and Gilles Seutin asked scientists from developing countries to summarize their experiences of international collaboration and to suggest attitudes and practices that would lead to more fruitful exchanges with northern scientists. They also asked scholars to provide an analytical framework in which these issues could be discussed and to identify possible solutions to questions such as: What are the responsibilities of first world scientists involved in conservation actions in developing countries? How can biologists work toward the protection of biodiversity while being respectful of the human desire for a better future? The resulting papers analyse specific situations encountered in countries such as Democratic Republic of Congo, Madagascar, India, and Panama and discuss the philosophical basis for environmental research. They also examine the work of two institutions whose projects in developing countries have been particularly effective through outreach and attention to local values and needs and who propose a pluralistic view of conservation biology ethics.

*Protecting Biodiversity* seeks to encourage students and professionals involved in conservation projects to adopt culturally sensitive attitudes that will lead to greater effectiveness and efficiency in conservation and greater respect for the differences of others.

CATHERINE POTVIN is associate professor of biology at McGill University. MARGARET KRAENZEL has recently completed her master's degree in biology at McGill University. GILLES SEUTIN is senior biologist in the Ecological Integrity Branch of Parks Canada where he coordinates the Species at Risk Program.

*To Michael Shenstone, who helped us from the very beginning and believed in our quest. This book would not exist without him.*

# Contents

# Acknowledgments

This book is the by-product of a symposium held during the 1996 IUCN World Conservation Congress in Montreal. The symposium was a cultural experience in itself, as original contributions were written in three languages: French, English, and Spanish. We greatly appreciate the support given to us by CIDA (Canadian International Development Agency). The Biodiversity Office of Environment Canada covered the travel costs of some authors and later provided English translations. We thus acknowledge John Herity of Environment Canada for the financial support that his office provided.

# Introduction

CATHERINE POTVIN AND
GILLES SEUTIN

Concerns over the risk of large-scale losses of biological diversity as a result of human activity were instrumental in the emergence of the field of conservation biology. This new area of research aims to develop tools that protect the Earth's diversity of genetic material, living organisms, communities, and ecosystems. As biologists take on the task of conservation, governments, international agencies, and non-governmental organizations have become increasingly pre-occupied with both the conservation and sustainable use of biodi-versity. The first two objectives of the Convention on Biological Diversity (CBD), which has now been ratified by 175 countries, are "the conservation of biological diversity [and] the sustainable use of its components."[1] Conserving and using natural resources are typically difficult goals to reconcile, however. Thus, conservation biologists often find themselves directly involved in politically and socially loaded issues that are made even more complex by cultural differences in values and perceptions.

Several years ago, work opportunities brought us to southern countries where we discovered a completely different environment. It soon became apparent to us that the protection of the extraordi-nary diversity of life forms in the South could not be divorced from local people's use of their environment's resources for their subsis-tence. Thus, our concern for the tropical environment became entangled with a concern for human beings. We started asking ourselves the fundamental questions that generations of philosophers have entertained: Should nature be used? How? How much? Our formal training in traditional biological disciplines – namely plant ecology and evolutionary genetics – had not prepared us for this questioning.

It became clear to us that the biodiversity crisis, and possible solutions to it, can be traced back to our understanding of the human/nature duet. Different societies have different understandings of the place of humans in nature. For instance, the notion of wilderness – untouched and inaccessible landscapes – is prevalent in North America.[2] Ecosystems in Europe have been so intensely modified by humans, however, that wilderness of this kind is nowhere to be found.[3] In "The Roots of Heaven" (1958) Romain Gary vividly portrayed how two cultures may have different relationships to the same animal: "Pour l'homme blanc, l'éléphant avait été pendant longtemps uniquement de l'ivoire et pour l'homme noir, il était uniquement de la viande, la plus abondante quantité de viande qu'un coup heureux de sagaie empoisonnée pût lui procurer. L'idée de la 'beauté' de l'éléphant, de la 'noblesse' de l'éléphant, c'était une idée d'homme rassasié."[4] Efficient actions for the conservation of species or habitats thus require a solid understanding of the needs and value systems of the people who exploit or otherwise benefit from the existence of that resource.

François Ost[5] identified what may be an even greater problem when he claimed that contemporary European and North American citizens have lost the sense of their link with, and their distance from, nature. Modernity has objectified nature, but has proven itself insufficient to protect the environment.[6] As a result, new philosophies have emerged from the ideas of Aldo Leopold and others. These philosophies transform nature into a sacred grove. Norton, Ost, and Larrère and Larrère all argue for a new environmental ethics to challenge these opposing views. They call for a profound change in our current environmental ethics, in order to bring about significant changes in lifestyle.[7]

The intent of this book is neither to take sides in this debate, nor to challenge specific views. Ariadne's thread of this book lies in our search, as practitioners of conservation biology, for answers to the difficult ethical questions that we faced in the field. Soon after beginning our work in the tropics, we started to question our attitudes in performing research. We began to wonder whether conser-

vation actions might sometimes constitute a new form of imperialism, a theory that has been proposed by scientists from Southern
countries.[8] We started questioning the way in which scientists from
Western countries conduct their work in Southern countries. Something seemed terribly wrong in the stories that reached us about the
actions of some conservation biologists. For example, we heard
about an American scientist fighting with local fishermen in Peru to
throw the fishermen's harvest of marine turtles back into the sea.
This story may or may not be true; nevertheless, it launched us on
a quest for guidance in our professional actions. We searched conservation biology journals, looking for the code of ethics of conservation societies. We found a three-page editorial in *Conservation
Biology* that considered ethical issues.[9] While the proposed code of
ethics was valuable, it was not accompanied by the in-depth discussions of rationales that we were looking for. Textbooks in conservation biology did not provide much guidance either. Very
recently, the Biodiversity and Ethics Working Group of the Pew
Conservation Fellows published the results of a broad survey of
biodiversity research protocols.[10] Their conclusions confirmed our
impression that the ethical dimension of conservation work is not
appropriately addressed: out of the 226 conservation organizations
that were surveyed, 75% had no ethical guidelines.

Throughout this book, we strive to determine what "good practices" may be for conservation biologists. We explore current philosophical quests for new environmental ethics[11] and attempt to
apply such morality to field practices in foreign countries. The book
draws from the experiences of Southern conservation biologists and
the theoretical constructs of social scientists. It presents suggestions
aimed at mobilizing conservation biologists to establish rules that
would ensure "good practice" in the field.

THE BOOK'S ORGANIZATION

Conservation work in developing countries is often driven by priorities that are set by donor countries or individual scientists from
the northern hemisphere. Too often, little attention is paid to the
perceived areas of need for such work by national authorities or
local communities, and to suggestions they make on ways in which
to conduct it. In this book, we reverse this usual way of doing
things. We have asked contributors from developing countries

to identify i) the needs of biologists and resource managers from Southern countries who are involved in the protection of biological diversity; and ii) the attitudes that foreign scientists should adopt when they work abroad. We then ask philosophers, anthropologists, and development experts to develop explicit theoretical concepts that address the suggestions made by the practitioners. All authors explore the diverging perceptions that Western and Southern scientists have of their endeavours to protect biological diversity, and how these may limit fruitful collaborations. We hope that the suggestions provided here will stimulate greater, more open, and more equal exchanges between partners in conservation actions and lead to a better understanding and sharing of respective goals. We see this volume as a step towards establishing an open dialogue between conservation researchers and practitioners from Southern and Western countries.

We have structured *Protecting Biological Diversity: Roles and Responsibilities* in the form of four dialogues between authors. In the first dialogue, Léonard Mubalama (chapter 1) tells us of the bad experiences he has had in the field, due mostly to the "poor practices" of foreign conservation biologists. This discussion opens the door for Marie-Hélène Parizeau (chapter 2), who lays the foundation of what could be a code of ethics for conservation biologists. The second dialogue, between Rogelio Cansarí (chapter 3) and Priscilla Weeks and co-authors (chapter 4), centres on attitudes and perceptions. It emphasizes the need for conservation biologists to understand their own culture, and the one in which they are working, in order to make conservation action both just and effective. Lala Rakotovao and co-authors (chapter 5) emphasize that conservation action in Madagascar has been impeded by the notion that biological diversity should be conserved for its own sake. She refers to this as a "misconception." In answer, Bryan Norton (chapter 6) examines the polarization that underlies the valuation of biodiversity. Rather than using the dialogue form, chapters 7 and 8 provide useful tools to help improve the effectiveness of conservation action.

A fundamental axiom of this book is that conservation biologists are scientists working in the social arena. Consequently, they must be equipped with strong social skills and must clarify the ethical underpinning of their actions. We hope that this book will contribute to better practice and collaboration among national and foreign scientists and local populations. Putting together this book taught us that cultural diversity is strongly coupled with biologicial diversity. We believe that failure to recognize this relationship will prevent the successful preservation of biological diversity.

NOTES

1  Convention on Biodiversity. *Text and Annexes.* UNEP/CBD/94/1, 94-04228. April 1998.
2  R. Nash, *Wilderness and the American Mind* (3rd ed.) (New Haven, CT: Yale Univ. Press, 1982).
3  C. Larrère and R. Larrère, *Du bon usage de la nature. Pour une philosophie de l'environnement* (Paris: Alto Aubier, 1997).
4  Romain Gary. *Les racines du ciel* (Gallimard, France: Collection Folio, 1983).
5  F. Ost, *La nature hors la loi. L'écologie à l'épreuve du droit* (Paris: Editions La Découverte, 1995).
6  C. Larrère and R. Larrère, *Du bon usage de la nature. Pour une philosophie de l'environnement* (Paris: Alto Aubier, 1997).
7  D.W. Orr, *Ecological Literacy. Education and the Transition to a Postmodern World* (Albany, NY: SUNY Press 1992).
8  W. Sachs, *Gobal Ecology* (London, New Jersey: Zed Books, 1993).
9  J.G. Colvin, "A Code of Ethics for Research in the Third World," *Conservation Biology* 6 (1992): 309–11.
10  T. Churcher, *Biodiversity Research Protocol* (Ernets Orlando Lawrence Berkeley National Laboratory, 1997).
11  C. Larrère and R. Larrère, *Du bon usage de la nature*; Orr, *Ecological Literacy* and B. Norton, *Towards Unity Among Environmentalists* (NY: Oxford University Press, 1991).

*Protecting Biological Diversity*

# 1 The Development and Management of Protected Tropical Areas: The Need for a Code of Ethics to Guide Collaborative Research in Africa
## A Case Study from the République démocratique du Congo

LÉONARD MUBALAMA

### EDITORS' INTRODUCTION

Léonard Mubalama obtained his Master's degree at the Durell Institute in the United Kingdom, where he examined the protection of African elephants. Until recently, he worked at Ituri National Park in Zaïre, but political tensions and armed conflicts in the Great Lakes region forced him to leave for Tanzania. His contribution summarizes obstacles to the protection of biological diversity in his country, as he recounts personal experiences of international collaboration tainted with inequity. Above all, Mr Mubalama calls for respect in international collaborations, through respect for local scientists and local populations.

### INTRODUCTION

To date, ecological studies have identified and described some 1.7 million species on the earth, but it is estimated that there likely exist between 5 and 100 million species, of which the vast majority remain to be described.[1] In terms of biodiversity, the République démocratique du Congo (formerly Zaïre) is the most important country in Africa, and one of the most widely known in the world.[2] It ranks third in the world after Brazil and Indonesia, and contains 9 per cent of the world's rainforests. It is ranked fourth in mammalian diversity after Indonesia, Mexico, and Brazil, with 409 species. Furthermore, the country is abundant in endemic species, notably the okapi (*Okapi johnstonia*), the Congo peacock (*Afropavo congensis*), the Pygmy chimpanzee (*Pan paniscus*), and the northern white rhinoceros (*Ceratotherium simum cottoni*), of which only twenty-nine individuals remain on the entire planet. All of these are to be found in the nature sanctuary at Garamba National Park.

As well as possessing great biodiversity and remarkable endemism, the République has 1,086 known species of birds, 216 species of amphibians, 48 species of butterflies, and approximately 11,000 species of plants. In light of the increasing number of scientific explorations taking place in this area, these numbers are likely to grow.[3] The encouragement of these explorations was one of the objectives of the Institut Zaïrois pour la conservation de la nature (IZCN), which sought to promote basic and applied research focusing on the efficient management of the country's protected areas.

In spite of the country's remarkable biodiversity and the obvious interest shown by foreign and African biologists in learning about it and in safeguarding it, to date little attention has been paid to the relationship between the needs of the local populations and foreign biologists' attitudes when working in the field. Obvious problems are associated with the professional code of ethics of foreign biologists, who seem increasingly interested in carrying out their research in "natural laboratories," such as Africa's national parks and protected areas.

An improvement in the attitude of foreign biologists working in Africa would help scientists meet the needs and visions of the local communities.

DEVELOPMENT

Because of the many factors that contribute to inefficient management, particularly the lack of trained scientific personnel, state-controlled services responsible for the management of Africa's protected areas are increasingly soliciting logistical and financial support, and technical expertise, from non-governmental organizations (NGOs), as well as support through bilateral and multilateral partnerships. This form of assistance has contributed to and continues to ensure the conservation of the protected areas in most African countries.

It goes without saying that expertise in scientific research is mostly found in Western countries, and particularly in North America. Only six per cent of qualified scientists and technicians are found in the southern hemisphere;[4] hence the need for African countries to seek out Western expertise whenever they face urgent nature conservation issues.

In the République démocratique du Congo, and more specifically in Virunga National Park (the first national park created in Africa, in 1925, it comprises 8,000 square kilometres), co-operation with NGOs has been one of the major factors contributing to the proper management of the ecosystem. Many scientific studies have been done within this particular framework, including research on the ecology of the hippopotamus (*Hippopotamus amphibius, L.*) and that of the warthog (*Potamochoerus porcus*). Recently, studies on the habitat of the mountain gorilla (*Gorilla gorilla beringet*) and chimpanzee (*Pan troglodytes schweinfurthii*) have been conducted. This study not only provided the park with a database for effective management, but also allowed for the training of field personnel.

Almost six years ago, CEFRECOF (Centre de formation et de recherche en conservation forestière), was created in the Ituri forest at the Okapis Wildlife Reserve (Reserve forestière de faune à Okapis – RFO). The centre – a product of close collaboration between the New York Zoological Society, a division of the Wildlife Conservation Society (WCS), the United States Agency for International Development (USAID), and the Institut Zaïrois pour la conservation de la nature (IZCN) – developed a permanent training program for its personnel and for the faculty of science at the University of Kisangani. Thanks to the field research of the WCS, in collaboration with IZCN, the Gillman Investment Company (GIC), and the World Wildlife Fund (WWF), the Okapis Wildlife Reserve management plan was developed – the first of its kind created for a protected area in the country. Furthermore, for the past two years CEFRECOF (Centre de formation et de recherche en conservation forestière) has been involved in an ambitious training program

for IZCN's qualified researchers, all within the co-operative framework developed between the WCS and the Durrell Institute of Conservation and Ecology (DICE) of the University of Kent in the
United Kingdom. This post-graduate study program is aimed at
improving the development and management skills of IZCN's scientific personnel.

Although, overall, these co-operative and collaborative projects
have been remarkably successful, and in many ways are producing
tangible and satisfactory results, much remains to be done in developing and improving the relationship between the needs of local
populations and the behaviour of foreign researcher-biologists in
the field. In this new century, ethics is becoming one of the most
pressing concerns in the field of biological conservation. It is
becoming more commonly understood that good management
policy for national parks and protected areas stems from thorough
knowledge, appropriate development plans, and the monitoring of
plant and wildlife resources.

In the past, discourteous behaviour and negative attitudes have
sullied relationships in the field or in local communities between
foreign experts and their native counterparts. The Lulimbi scientific
station in Virunga National Park serves as a good example of a site
of tense relationships between collaborators from different hemi-
spheres. In the 1970s, this African station was established with the
help of Belgian technical co-operation. The primary objective of the
project was to promote scientific research on the efficient management of Savannah ecosystems. In 1972–73, this station was frequently the scene of tumultuous confrontations between the Belgian
team of co-operating researchers and their African counterparts.

The members of the Belgian-African team had opposing views on
issues ranging from logistics management to proper jurisdictions, to
the point where top management at IZCN was frequently called
upon to settle disputes and to alleviate the endless tensions between
the two camps. A well-known incident involved an African
researcher who was refused the right to take his seriously ill spouse
to the health centre, which was located about 32 kilometres from
the station, while the Belgian expert allowed himself the luxury of
using the vehicle to spend the weekend in the regional headquar-
ters, which was located almost 150 kilometres away!

Furthermore, many cases in which foreign researchers working in
this same park ignored their contractual obligations were remarked

upon and deplored. What can be said of the leader of a team of foreign researchers who, at the end of his term, violated IZCN protocol and took with him all the scientific samples, knowing full well that the collection and handling of such samples was bound by pre-established terms? This action was a direct infringement of the laws governing the protected area, a World Heritage Site.

The arrogant and shocking attitudes and behaviour of some European experts have been denounced by their African counterparts as neo-colonialist. For example, a European project leader, while clarifying a point during a meeting with his colleagues, plainly stated that "I am paying you a monthly collaboration fee far superior to what your government is offering you, so you have no right to complain. In fact, you should consider yourselves lucky."

No more than six years ago, a study on the local population of hippopotami (*Hippopotamus amphibius, L.*) was carried out in Virunga National Park. As usual, the Institute appointed me, a local researcher, to help carry out the work. After a tremendous amount of effort, performed in collaboration with the expert representing the European Economic Community (EEC), upon release of the final report on our co-operative efforts in the field I was surprised to see that I had been ignored as co-author of the document. This behaviour was clearly scientific imperialism – our Institute had been assigned to monitor this project in the absence of my colleague. Naturally, such behaviour was not likely to encourage our help.

These examples illustrate behaviour that is unacceptable to African researchers. Such acts hamper important collaborative advances in the field, including the conservation of ecological processes, life-sustaining systems, and the sustainable use of African species and ecosystems.

Considering the current trend throughout the development of an alliance movement of indigenous peoples and rainforest tribes, which was initiated at a conference held in Penang, Malaysia in 1992, local communities have started wondering if they are actually collaborators in the scientific expeditions done on their land or if they are simply the object of studies performed by foreign researchers. These communities have an increasing tendency to accuse managers of the protected areas of maintaining "human zoos to satisfy the curiosity of scientists and tourists."[5]

As a result of such unfair behaviour, several African countries have gradually begun implementing restrictive measures that will reinforce compliance with research legislation in the field of nature conservation. Local communities living alongside protected areas have filed many grievances against foreign researcher-biologists regarding attitudes expressed in the field. The most common complaint from locals is that these researchers are paying less and less attention to the sociopolitical and economic support that locals depend on as compensation for their traditional rights, rights that are infringed upon by the creation of protected areas on their land. Article 42 of the "Alliance mondiale des peuples de la forêt tropicale" calls to our attention: "The best guarantee of the conservation of biodiversity is that those who advocate it also uphold our rights to the use, the administration, the management, and the control of our lands. We affirm that the protection of the various ecosystems must be entrusted to us, the native peoples, given that we have lived on this land for thousands of years and our very survival depends on it."[6]

As well, native peoples who willingly participate in surveys on flora and fauna rarely receive adequate financial remuneration.[7] Furthermore, in spite of the fact that natives have proven that they are the best protectors of the very biodiversity that researchers attempt to discover and eventually to protect, few researchers have helped local people in the battle for ownership of the land. The equation becomes more complicated once the local population realizes that not only do they not have access to the results published in the studies in which they have often actively participated but that they are completely left out of the decision-making process of the management of protected areas. For instance, the first copies of the Okapis Nature Reserve Management Plan, which was sent to managers and other partners of the reserve, were written solely in English, while the staff who were to execute the plan were francophones.

In the République démocratique du Congo, we have often heard certain (unpopular) expert biologists promise miracles so that their contracts would be supported by the local population and renewed by the government. In many cases, and much to the dismay of that population, these promises were not kept. Such dishonesty makes local peoples increasingly reluctant to support similar conservation projects in the future.

The problems related to the effective collaboration between foreign researchers and their host counterparts, and foreign biologists and local communities living in and around protected areas, is drawing more and more attention from the international scientific community. It is imperative that in the very near future the World Conservation Union (IUCN, or Union internationale pour la conservation de la nature et de ses ressources) examine and take into consideration the recommendations that I propose.

### RECOMMENDATIONS AND CONCLUSION

On a global scale, the major obstacles to the efficient management of protected areas in developing countries are the absence of adequate equipment, the lack of management tools (such as the development plan), and an insufficient number of qualified scientific personnel and wardens to protect an area to IUCN standards.

It is becoming increasingly obvious that, in addition to the factors mentioned above, the lack of a scientific code of ethics is becoming one of the basic impediments of any plan to ensure proper rational management of protected areas in Africa. It is imperative that all research protocols allowing researchers access to protected areas be revised by explicitly adding the terms of a scientific code of ethics. This would promote a new, co-operative relationship between foreign biologists and local populations. A code of conduct, aimed at bringing together conservation biologists and the public by reinforcing mutual respect and the sharing of knowledge and resources,[8] should be based on the following points:

1 Respect for local customs should be expected and enforced. In addition to obtaining official research authorization, it would be advisable that expert biologists who are planning a field study either in or surrounding a protected area contact the appropriate local authorities to explain the purpose of the study. This would eliminate any unpleasant surprises, which might upset the progress of the project. It would also be useful for foreign biologists to familiarize themselves with a guide outlining the socioeconomic situations of the communities that may be affected by legislation on protected areas. Such studies should be implemented by Africa's protected areas, and should show the ethnic diversity of the communities and their social structures, in order to give foreign experts in-depth knowledge of local populations so as to avoid any possible misunderstandings.

2 Clearance from the appropriate central or state government authority and compliance with the legislation controlling the protected areas must be explicitly maintained throughout the entire period of work.

3 Co-operation between foreign biologists and their local counterparts must be such that the latter will benefit from adequate training, which, in turn, will qualify them to continue monitoring activities at the end of the expert's term. Scientific knowledge is to be shared.[9]

4 It is advisable that foreign biologists working in the field refer as much as possible to the collaborative approach of the Participatory Rural Assessment (PRA), which promotes full integration of Native peoples into the execution of research projects and ensuing discussions about the way in which results can be used to benefit the population, rather than treating them as simple objects of investigation. This policy is presently being implemented at CEFRECOF at the Okapis Wildlife Reserve, where Natives are involved in conducting wildlife and nature surveys. It should be implemented on a larger scale.

5 Native colleagues' copyrights must be recognized when the results of co-operative field studies are published, whenever the copyright rule applies.

6 Foreign researchers must publish the results of their work and make them available to the national conservation agency of the host country in its official language, and within a reasonable period of time.

7 Foreign biologists should avoid making false promises to local communities; on the contrary, surprises such as the building of a health centre to benefit the community would be greatly appreciated and would encourage the community to be more co-operative with the managers of protected areas.

8 Once their field work is completed, conservation biologists working in the southern hemisphere should spark local populations' interests at both non-governmental and governmental levels, to ensure that locals play an integral part in planning and implementing management plans for protected areas.

The fulfillment of such a code of ethics requires a consistent commitment to conservation by the host country, as well as compliance with the terms of research convention or protocol. Perhaps a similar code of ethics should apply to the host country as well.

NOTES

When Mr Mubalama came to Montréal for the 1996 World Conservation Congress he left Zaïre. When he returned, war had broken out, and Zaïre was no more. This political turmoil is responsible for our reference to "l'Institut Zaïrois pour la conservation de la nature."

1  United Nations Environment Program (UNEP), *Global Biodiversity* (Nairobi: UNEP/GEMS Environment Library No.11, 1993).

2  S.N. Stuart and R.J. Adams, *Biodiversity in Sub-Saharan Africa and Its Islands* (Oxford: Occasional Papers of the IUCN Species Survival Commission No.6, 1990).

3  J.A. McNeely, K.R. Miller, W.V. Reid, R.A. Mittermeier, and T.B. Werner, *Conserving the World's Biological Diversity* (Gland: The World Bank, World Resources Institute, World Conservation Union, Conservation International, and World Wildlife Fund, 1990).

4  National Academy of Sciences, Committee on Research Priorities, *Research Priorities in Tropical Biology* (Washington, DC: National Academy of Sciences, 1980).

5  J. McKinnon, K. Child, and J. Thorsell, *Aménagement et gestion des aires protégées tropicales* (Gland: IUCN, 1990).

6  FTP Newsletter, "Global Alliance of Indigenous People of the Rainforests," 18 September 1992.

7  J. Clay, *Indigenous Peoples and Tropical Forests – Models of Land Use and Management from Latin America* (Cambridge, Mass.: Cultural Survival Ind., 1988).

8  J.G. Colvin, "A Code of Ethics for Research in the Third World," *Conservation Biology* 6, no.3 (1992): 309–11.

9  Ibid.

## 2 Considerations on a Code of Ethics for Conservation Biologists

MARIE-HÉLÈNE PARIZEAU

### EDITORS' INTRODUCTION

The frustrating experiences related by Léonard Mubalama in chapter 1 are neither unique nor exceptional. We are inclined to think that they are often the incidental result of well-intended initiatives led by poorly informed people. One avenue leading to a resolution of this problem is the elaboration of a code of ethics for conservation biologists engaged in international actions. In this chapter, Marie-Hélène Parizeau, a philosopher at Université Laval in Quebec, examines the postulates of conservation biology and demonstrates that it cannot be a neutral science, since it is based on the assumption that biodiversity is "good." Dr Parizeau then uses the field of medicine as a model to outline a possible code of ethics for conservation biologists.

### INTRODUCTION

Biological conservation, defined as "the study and the protection of biological diversity in the context of ecological crisis,"[1] is a relatively new discipline, first recognized by the scientific community in the 1980s. Through textbooks, specialized papers, and fieldwork, it is now being taught in Canadian and U.S. universities. Conservation biology proposes specific models of intervention in the environment in order to conserve or reinstate the biological diversity of given areas that are inhabited to a greater or lesser degree by human communities. These interventions are not neutral actions, as they tend to modify not only the environment but have an impact on the human inhabitants as well. They therefore demand, on the part of the conservation biologist, ethical reflection concerning his or her goals and the methods by which he or she plans to meet them. In this part I analyze the necessary premises of a code of ethics for conservation biology. First, I will critically examine the postulates and the characteristics that form conservation biology, to lay bare its underlying values. Second, since conservation biology is becoming a recognized professional subject, I will show that this discipline must be ethical in meeting its ends. Finally, I will determine the limits involved in the ethics of biological conservation by isolating specific leading principles of the discipline, based on the naturalist paradigm and the metaphor of the "doctor of nature," which both lead to problems.

### CONSERVATION BIOLOGY: A CRITICAL ANALYSIS OF ITS POSTULATES

As a scientific discipline, conservation biology has its own postulates and an accepted view of intervention, which include demonstrating both the values that must be defended and the ways in which conservation biology professionals must behave. Let us critically examine the four main postulates and characteristics of conservation biology, as discussed by Meffe and Carroll (1994).

> *1 "The diversity of living organisms and of ecosystems is inherently good."*[2]

This statement is both a scientific statement (a system that lacks a certain degree of diversity is weak one or in peril), and a value judgment (there is a moral obligation on the part of the biologist to promote biological diversity). This value judgment has a corollary: biological uniformity is bad because it goes against natural laws.[3] It is important to note that one cannot simply slide derive an "ought" from an "is"; there is a class of statements of fact that is logically distinct from a class of statement of value.[4] In this statement there is a moral, even ontological postulate, involved right from the beginning. This postulate states that "we human beings have a moral obligation toward ourselves and toward future generations; to maintain and promote the conditions that are necessary for biological and human existence."

## 2  *"Biological complexity is good."*[5]

Stating this implies that the biologist has made a value judgment that what is natural or "wild" is intrinsically better than what is artificial. Thus, in the Convention on Biological Diversity (CBD), the biologist prefers *in situ* measures to *ex situ* ones. This postulate also carries a "nature knows best" feeling – a belief that nature knows how to take care of itself. The biologist therefore has an obligation to respect the means by which this natural complexity is expressed. It is worth noting that the distinction between wild and artificial is of epistemological relevance for the conservation biologist and we should ask if it is pertinent in any case in which we believe that a technical (artificial) intervention on nature by the biologist contradicts the natural means of creating biological complexity. In other words, we must question a) the status of the technical interventions by the conservation biologist; and b) the particular representation or historical construction of the relation of nature used in biological conservation, which is a scientific discipline put forth by Western society, particularly by Americans.[6] For example, certain groups insist on the concept of "wilderness," which can mean many things for many people. In Europe, "wilderness" does not mean anything biologically significant. Instead, the concept of natural heritage[7] is preferred, because wild forests no longer exist and human culture has significantly shaped the landscape.

## 3  *"Evolution is good."*[8]

Viewed as a natural phenomenon, evolution serves as a guide for possible interventions by biologists. This is a fact-based judgment (evolution helps maintain and develop the conditions necessary for biological life), as well as a value judgment (we must follow natural evolution). An important question to be considered is whether human beings and their technical manipulation of nature are a part of natural evolution or are external to it. Should conservation biology include the transformation of human societies as part of an evolutionary paradigm and, if so, under what conditions? Asking this question acknowledges the complicated problem of the relation of man to nature, as well as the presence of innumerable theories, ranging from sociobiology to moral naturalism to ecocentrism.[9]

### 4 *"Biological diversity possesses an intrinsic value."*[10]

This strictly moral statement puts forth the idea that nature cannot be viewed solely as a means to human ends. In other words, under certain circumstances, nature's interests in self-conservation are more important than humans' utilitarian purposes. The problem thus resides in establishing moral decision-making criteria that would enable us to decide how to share resources fairly between human beings and other living species. Unfortunately, stating the ethical problem in such a way (as to give an ontological meaning to biological diversity) may not be a solution. Bryan Norton (see chapter 6) provides a critical analysis of value theory within this context, and his adaptational model could be one substitute, among others, for this last postulate. It is important to decide how to deal with this problem, as it is the core of Article 1 of the CBD.

Some further characteristics of conservation biology should be added to these starting postulates in order to better define it.

### *Multidisciplinarity is necessary.*[11]

Evolution in biological conservation must take into account and integrate factors that are anthropological, sociological, economic, and political. One cannot completely separate nature (the environment) and human beings: their relation must be incorporated into any theory of conservation biology.[12] Attempting to articulate the relationship between nature and human culture is an evolving epistemological problem.

### *The necessity to act exists within a climate of scientific incertitude.*

In environmental matters, biologists have not completed the inventories that would enable them to make scientifically sound decisions. The knowledge we now possess is fragmentary and must be understood in context (time and space). In the moral realm, this scientific uncertainty gives rise to two different attitudes – caution and pragmatism. Caution forces us to measure the consequences of an action and to act reasonably and modestly. Pragmatism implies that we must remain attentive to additional factors and new data while maintaining the same goals (e.g., the protection of biodiversity).

> *Vouching for the protection of biological diversity is not neutral; it demands a professional attitude that goes beyond a simple militant one.*

For the conservation biologist, a delicate moral problem is finding a way in which to communicate scientific data to the public as well as those in power politically, without actively engaging in a militant movement.

Conservation biology is thus a scientific discipline that contains the imperative for concrete action leading to a precise end – the protection of biological diversity in order to maintain the possibility of human and other biological lives in the future – and can be opposed to many current human interests – particularly technological and economic growth.

## CONSERVATION BIOLOGY AS A PROFESSION

Conservation biology, viewed as a new discipline for technical intervention in the environment (intervention for the preservation of biodiversity), can also be viewed in light of the professionalism movement that arose in post-industrial, twentieth-century society. "The concept of professionalism refers to persons or groups of persons that are specialized according to their different goals, such as the acquisition of theoretical and practical knowledge necessary to work in a specific domain of social division and technical labour. These persons progressively assimilate the values privileged by the occupational collectivity of which they are a part and adopt attitudes and behaviours that must be followed in the exercise of their professional activity."[13]

What link may be established between professionalism and biological conservation? An occupational activity that is socially recognized implies a certain amount of theoretical and practical knowledge, which reflects certain values that must be translated into specific behaviours. In the first section, I have described the principal beliefs and characteristics of this new discipline. These serve as practical indicators as well as behavioural values to be followed by the conservation biologist. As a biologist doing everyday work, one must ask oneself, Am I doing the right thing? Technical actions are not neutral, and the knowledge acquired may have repercussions on many levels – individual, collective, political, etc. For example, whose well-being was being taken into consideration when Native collectives were forced out of sites dedicated to the establishment of natural parks? Should a biologist decide to release turtles from an unaware fisher's net?

The ethical questions What should I do? or What is the right thing to do? are at the very heart of our professional, personal, and social activities. In everyday life, we speak of good, evil, justice, moral obligation, duty, responsibility of persons, of moral community – a vast ethical vocabulary with which we are more or less familiar – and that is used in our usual conversations and behaviours, without our necessarily being aware or explicit. "He lied to me"; "She is not professional"; "This corporation has forced unacceptable and unjust work conditions"; "That program does not respect equality criteria"; all of these statements contain value judgments linked with our professional, social, and private activities. Because of our personal and family history, we are all in possession of a more or less coherent and continuous internalized, constructed, or adopted social and historical context from which we take values that provide meaning and guiding principles throughout our lives. These are called personal ethics (the ancient Greeks, Aristotle in particular, spoke of the ethics of a good life when referring to the means necessary to the attainment of a virtuous life).

Personal ethics pertain to each individual's private life. There is also the parallel existence of a social ethic, which gathers the consensual values adopted by a given society and which may be translated, in Western society, into a judicial system that ensures that these consensual values are effectively respected and followed. For example, in Canada, and in Quebec, this "core" of values is expressed in the Canadian Charter of Human Rights, which was a direct descendant of the Universal Declaration of Human Rights, enacted in 1949. Every law put forth by civil society must be in accordance with what is contained in that charter of rights. Through this concept of justice, we can see that ethics and law form a pair, even though they do not serve the same functions. Ethics rely on ideals and demand individual internal conviction; law is used to settle conflicts between individuals and social groups and to express limits that must not be crossed.

Certain types of social activities (traditionally, medicine and law) have established quasi-juridical rules to maintain values for their professions. These are professional codes of ethics, or deontological codes, that provide the just rules of conduct that professionals must follow, as well as sanctions for those who deviate from them. Deontological codes are tools of social regulation, which have multiplied at the same rate at which our Western technological civilization has developed. These deontological codes and rules of conduct contain values that are at the very basis of a profession. They also contain rules of action that evolve through time, so that they can be of use when a professional needs to make an ethically sound decision in situations in which the factors involved make it difficult to determine what conduct should take place.

Unlike medicine, conservation biology does not possess whole books (such as international codes of ethics, treatises, or deontological codes dating as far back as the fifth century BC under Hippocrates) that attest to an evolving ethical questioning taking place through history. In biological conservation, this ethical reflection has yet to occur, although the need for a professional ethic is very real. In the first chapter, Léonard Mubalama presents a number of circumstances in which a code of ethics would be beneficial. These circumstances vary widely in scope, ranging from conflict over the use of field vehicles to the illegal export of samples and the lack of proper credit for intellectual contributions. (The circumstances he discusses mainly pertain to relationships between foreign and national scientists.)

Because conservation biologists work with local communities, ethical guidelines are also needed to guide the scientist/community relationship. In the next chapter, Rogelio Cansarí discusses specific ethical dilemmas relevant to scientists working with indigenous people. He emphasizes the mistrust that indigenous people have for foreign scientists because of the scientists' perceived duplicity, lies, and failure to comply with agreements. His comments draw attention to another realm of ethical issues: those guiding the interactions between scientists and local populations.

Because it is so evident that a code is needed, I will attempt to put forth a code of conduct for conservation biologists.

### SOME STANDARDS FOR AN ETHIC OF BIOLOGICAL CONSERVATION

The postulates and statements put forth in the first section of this paper are to be used as the basis of an ethic of biological conservation. From this ethic, it will be possible to establish practical guidelines that can be used by conservation biologists to delimit the nature of their interventions, to examine the underlying values of their actions and ideas, and to resolve any conflicts of values that may arise.

Following its postulates, it appears that conservation biology maintains a naturalistic vision of the relationship between humans and the world. Naturalism[14] signifies that, in its laws and ways of working, nature shows humans the ways in which life may perpetuate itself and follow its natural evolution. Moral wisdom consists in taking into account the mechanisms by which nature works because, ultimately, nature understands better than humans do what it needs in order to take care of itself.[15] In this article I advocate for a biological conservation based on a weak naturalism that sees morality in relation to nature, not in opposition to it, thus avoiding reductionism (human culture is different than biological order). We could speak of a certain type of complicity between the biological and moral norm. This type of paradigm can be seen as being opposed to the mechanical paradigm of modernity, which shapes humans out of their natural condition through the accelerated development of technology, science, and the idea of human progress.

*If* conservation biology wants to maintain[16] this naturalist paradigm, I propose that, to remain coherent, the conservation biologist must assume the role of nature's physician, by treating ecosystems that are now sick because of certain human activities. This analogy was also proposed by Michael Soulé in his 1985 work.[17] If the biologist perceives himself as nature's doctor, she or he will have to follow a certain medical reasoning, namely, diagnosis, decision as to treatment, and, finally, consent to carry out treatment. The first step, which is to propose a diagnosis, implies methodological precision to identify, itemize, systemize, and compare biological diversity and its clues, in order to better diagnose its problems. A diagnosis will always be made with some scientific uncertainty, a level of ignorance of greater or lesser importance, and an element of contextualization (that is, certain problems may not be as important as others). Both Mubalama (see chapter 1), and Cansarí (see chapter 3) express the desire that local and indigenous knowledge be taken into account in the diagnostic phase. To make a proper diagnosis means one must ask questions of the people concerned about whether they are responsible for the pollution or as a resource to protect biodiversity, or both.

The second step is to propose a treatment, along with other alternatives. For the biologist, there may be one best solution; pragmatically, however, in order to attain the assent of as many people as possible to a plausible solution, it is necessary to be flexible and to propose other alternatives, including ones that seem to be of lower quality. In chapter 4, Priscilla Weeks and her co-authors introduce the notion of cultural lenses and develop the idea that the perception of environmental problems and associated solutions are largely determined by cultural viewpoints. Thus, practising conservation biologists have to be conscious of their own cultural lenses and acknowledge that their counterparts within a given society, as well as the local population, may not share their opinion about diagnosis or treatment. In a concrete situation, the biologist must offer different rehabilitation measures – for example, the establishment of protected zones or the development of educative and political measures – and then, ideally, a consensus can be reached by all stakeholders.

The third step is gaining consent to begin treatment. Here we must acknowledge a very important ethical point: nature can never grant consent, but human beings have the power to either choose to conserve nature or to use its resources. There is no escape from anthropocentrism: human beings decide, speak, and act on nature's behalf. Consent to treatment (the modalities used for the conservation of nature) must be made by people – the conservation biologist is not alone, neither in his or her methodologies, which demand multidisciplinarity, nor in the ultimate decisional process, which leads to action that enables conservation.

Who are these people who give consent either to defend their own interests or to represent nature? Following local contexts, they are local communities, Non-Governmental Organizations (NGOs), governments, and international communities. Many potential ethical problems may surface, because each group has different interests. The concept of consent to treatment shows that the work of a conservation biologist continuously involves dialogue with other people or groups of people in order to determine the future of an environment shared by all. The conservation biologist cannot be only a technician who pinpoints, masters, and acts on a given piece of reality – his or her professional responsibility consists in proposing and executing collective actions that will lead to the preservation of biological diversity.

This third step may be the most challenging for conservation biologists. The notion of "consent to begin treatment" differs from the modern concept of expertise in that scientific knowledge is no longer seen as sufficient[18] – partnership is required if action is to succeed. This notion requires that conservation biologists understand and accept the limits of their knowledge and are open to the involvement of a broader community in decision making. Throughout this book, testimonies suggest that problems with local communities arose largely because the scientists failed to inform the community and seek consent vis-à-vis the proposed conservation action. As a consequence, conservation action was viewed negatively, and the biologists failed to gain popular support. Conversely, Butler and Regis (chapter 8) provide practical examples of various programs in which successful conservation endeavours are intimately linked to community involvement.

The professional model having been described, it is now possible to list some ethical principles that can guide the actions of the conservation biologist.

1  We cannot treat the environment without taking into account the presence of human beings (for example, communities inhabiting an environment).
2  Any act affecting the environment that is imposed in order to protect biological diversity consists of a human action and thus cannot be viewed as neutral, mainly because transformation of that environment and of the communities inhabiting it may occur.
3  Any act affecting the environment will lead to transformation; this act then possesses a power over the environment and the

human communities inhabiting it. This power must be wielded with respect for all people affected (national scientists, local communities, etc.) and any tendency to dominate (such as mastering, or objectifying) must be excluded. In order for this to be effective, people's dignity must be respected through the sharing of knowledge. In any other case, people are being considered as a means to an end and not an end in itself.[19] For example, this respect may be expressed in practice by teaching and training people from the local communities to put together measures to preserve or rehabilitate certain sites.

4 As in the therapeutic relationship, it is the duty of those who intervene to keep some matters confidential vis-à-vis third parties or outsiders (that is, to keep professional secrets). In the realm of biodiversity, certain areas of knowledge – ethnobotany, for example – may gain a monetary value, and thus are bound by the intellectual property copyright as put forth by judicial courts. Discretion is a moral imperative when economic stakes (e.g., the monetary value of certain forest species in a particular forest), or political stakes (e.g., one community's claim of political autonomy from another) are involved.

### CONCLUSION

These few principles open a new perspective on certain problems or ethical tensions that occur in the conservation biologist's work. One ethical tension inherent to the work of biological conservation is the struggle to put forth measures to protect biodiversity without ignoring the culture of the human communities that inhabit a given area, as well as their right to make use of its resources (see Weeks *et al.*, chapter 4).

The second tension is that of allegiance: on whom does the biologist depend? What are the political limits accompanying his or her mission? A very important ethical problem arises with the exploitation of knowledge of biodiversity in relation to globalization: what should the relationship be between the conservation biologist and the corporation, whose end is purely profit? In what situations could a biologist be found in a conflict of interest?

Another problem is that of colonialism, clearly illustrated by Léonard Mubalama, and Rogelio Cansarí (see chapters 1 and 3). For example, attitude problems often arise when a Western biologist is working in a Southern country, or when an urban biologist works in rural communities (see L.H. Rakatovao, chapter 5). How can we transform the relation of power into one of sociality, (for example, should the biologist behave as a guest does when invited somewhere)? Finally, a very important ethical problem is the introduction of technology into a less technologically developed society. How can we anticipate the consequences that such technologies may have on the organization of the social structure or the habits of that culture?

Questioning the ethics of conservation biology brings with it other questions – epistemological or professional deontological – and thus opens up yet more uncertainty for the practitioner. The will to protect biological diversity is not an ethical imperative that is shared by all individuals of the world, or by all human communities. Historically, it has been a Western concern. The fact that the international community has noticed, however, that there are problems concerning biodiversity and that it has organized conventions on that subject, suggest that the biologist has successfully stressed the urgency of his or her diagnosis and has been taken seriously. Nevertheless, the therapeutic treatments proposed have to meet the moral requirements of the protection of biodiversity and the human communities and cultures that make up that biodiversity, and also constitute a practical and philosophical response to the overexploitation of nature, which is the current model in Western countries.

NOTES

1   G. Meffe and R. Carroll, *Principles of Conservation Biology* (Sinderland, Mass.: Sinauer Associates Inc., 1994), 19–21.
2   Ibid.
3   The expression "natural law" is used in this text to remind us of the concept of biological normativity that dictates the essential conditions that must be met in order to have biological life at all, that is, intrinsic norms related to the phenomenon of life (for example, calorific energy, reproduction, genetic information, the cell as a basic structure, the diversity of forms, etc.). See D. Jacob, *La logique du vivant. Une histoire de l'hérédité* (Paris: Gallimard, 1970), 21–2.
4   John Searle, "How to Derive 'Ought' From 'Is'," in P. Foot (ed.), Theories of Ethics (London: Oxford University Press, 1967).
5   Here I refer to Hans Jonas, *The Imperative of Responsibility. In Search of an Ethic for the Technological Age* (Chicago: University of Chicago

Press, 1984), 47. The philosophical and ontological question is, Why ought something to be?

6   Meffe and Carroll, *Principles of Conservation Biology*, 19–21.

7   I am following ideas of the sociology of science, which calls scientific analyses of the world human inventions, dependent on anthropology, but I am also trying to shed some light on the conditions of possibility and validity, on which conservation biology is establishing its foundations.

8   See A. Micoud, "L'écologie de la mythe de la vie," in C. and R. Larrère (eds.), *La crise environnementale* (Paris: INRA Editions, 1997), 17–29; and A. Micoud, "En somme, cultiver tout le vivant," in M.-H. Parizeau (ed.), *La Biodiversité: Tout conserver ou tout exploiter?* (Bruxelles: DeBoeck, 1997).

9   Meffe and Carroll, *Principles of Conservation Biology*, 19–21.

10  P. Thompson's work gives a fair account of the problems at stake on this question. See P. Thompson (ed.), *Issues in Evolutionary Ethics* (New York: New York State University Press, 1995).

11  Meffe and Carroll, *Principles of Conservation Biology*, 19–21.

12  Ibid. All of these postulates are from Meffe and Carroll.

13  R. Larouche, *La Sociologie des Professions* (Quebec: Office des Professions, 1987).

14  See P. Acot, *L'histoire de l'écologie* (Paris: Presses Universitaires de France, 1988).

15  In moral philosophy, naturalism is understood as a "doctrine which stipulates that moral life is the perpetuation of biological life and that moral ideals are reflections of the needs and instincts that constitute our will to live .... It represents a philosophical tendency: anti-Christian, anti-Kantian, primacy of life and its perpetuation, the homogeneity of human and animal ends, optimism, to which we must add agnosticism and empiricism . . . . Thus conceived, naturalism is a doctrine which takes into high consideration the value of health, power, and of the survival of individuals. There also exist some theories, however, which transfer this value to that of health, power, and the survival of society, all viewed as a whole." A. Lalande, *Vocabulaire technique et critique de la philosophie* P.U.P. 1929, 16th edition, 1988.

16  This naturalistic paradigm was highly criticized in the history of biological philosophy. Some blame naturalism for the practice of quietism (the feeling that, if nature knows what's best for it, why interfere?); some, on the other hand, want to improve nature. As mentioned by A. Fagot, "In modernity, naturalism is less a philosophy than a polemic position. We must remind man of his modest position and condition ... Naturalism challenges both the very narrow view of anthropocentric humanism and the presumption that moral intuitions are transcendental. We are natural beings. We can neither hope for any radical difference from the rest of

nature, nor for a direct link with heaven . . . . For the naturalist, "genes hold our culture on its leash," therefore limiting the supposed contingency of our choices – cynicism or moral relativism. Nature guides us; we are within it. P. Changeux, *Fondements naturels de l'éthique* (Paris: Editions Odile Jacob, 1991), 193. Moral naturalism can be interpreted in strong or weak versions. This naturalist postulate deserves to be examined in a more systematic fashion than the simple critiques I have made in the first section of this paper. These postulates have considerable practical repercussions because they follow a cautious naturalistic ethic that helps limit the sort of economic and techno-scientific development that does not recognize the obligation to develop a long-term, durable development program for the environment and its local populations. In other words, such a paradigm proposes an ideal in which human societies are controlled in a strict fashion in order to maintain the conditions necessary for the diversity of life in nature.

17  Micheal Soulé, "What Is Conservation Biology?" *BioScience* 35 (1985): 727–34.

18  C. Larrère and R. Larrère, *Du bon usage de la nature: pour une philosophie de l'environnement* (Paris: Alto Aubier, 1997).

19  Here one must translate Kant's moral imperative "act in such a way that you treat humanity as well as you treat yourself and always treat it as an end in itself, never as a simple means."

## 3 The Scientific Community and the Indigenous Emberá Community of Panama

ROGELIO CANSARÍ

### EDITORS' INTRODUCTION

It has been fashionable to speak of indigenous peoples as the care-takers of the earth and to consider them the keepers of traditional ecological knowledge. Little is done, however, to help indigenous peoples look after their environments. Rogelio Cansarí, a recent M.SC. graduate in conservation biology from McGill University in Montréal, writes about the experience of his people, the Emberá of Panama, with anthropologists and, more recently, with bioprospec-tors. His testimony highlights the profound misunderstanding that exists between scientists and indigenous people, producing a situa-tion that leads to fear and potential clashes. He shows that scien-tists need to openly and clearly explain their research to people in communities in which they work. Indigenous Chief Bonarge Pacheco states the problem most poignantly: "If scientists do not tell us what they study, how can we help them?"

### INTRODUCTION

> These people [the scientists] come to our land and do anything they please; they learn from us, teach us nothing in return, and then we never hear from them again.

Worldwide interest on the part of many organizations, primarily international agencies, such as The United Nations Environment

*Panama: Location and Geographical Traits*

The Republic of Panamá is an isthmus in Central America, and forms the most narrow and, on average, the least elevated strip of land in the Americas. The country borders on the Caribbean Sea to the north, the Republic of Colombia to the east, the Pacific Ocean to the south, and the Republic of Costa Rica to the west. Panama's surface area is approximately 75,517 square kilometres. Panama is a mountainous country, with hills, valleys, and forests. The most important rivers are the Chucunaque and the Tuira, both of which are located in the Darién region, near the Colombian border. The location of Panama in the lower intertropical latitudes determines its tropical climate and vegetation; the country has relatively high temperatures that remain constant throughout the year and has both a dry and a rainy season.

Lowlands or warm-lands with altitudes below 1,420 metres cover 87 per cent of Panama; temperate lands with altitudes ranging between 1,420 and 3,040 metres cover 10 per cent of the country; and highlands or cold-lands, situated above 3,040 metres, cover 3 per cent of the country's surface area and are of volcanic origin. The Panamanian population reached 2,418,000 in 1990, 53 per cent of whom live in urban areas. The national population is mostly of mixed ancestry, as is the case in many other countries of Central and South America. Panama's indigenous population represents 8.34 per cent of the country's total population.[1]

Programme (UNEP) and the Convention on Biological Diversity (CBD); or environmental and other non-governmental organizations (NGOs); governmental organisations; and university, scientific, and student groups, to work with indigenous people has reached unprecedented proportions in the past decades. Aside from their unique cultural diversity, most indigenous people live in areas of great natural wealth that boast vast amounts of timber, oil, or mineral resources. For example, in Panama, the Kuna people live in a coastline forest along territorial waters that contain an incredible variety of marine resources and islands. Similarly, the Emberá inhabit one of the most biologically and geologically diverse forests in the country, and the Ngöbé-Buglé inhabit a very significant geological region, rich in mineral deposits. These areas are well-known around the world, and, in the face of growing interest, indigenous communities are adopting control measures over the visits to, and scientific research on, their cultures or territories. Very few benefits are reaped from this research by the indigenous people studied, however. Visiting researchers also display an astounding lack of professional ethics. In light of these facts, researchers need to adopt new strategies that improve present and future relationships with indigenous peoples and that redress past injustices.

Before any research activity is carried out by non-indigenous people on indigenous land, the procedures currently established in indigenous communities must be thoroughly understood. In this way, scientists will gain an understanding of the structures of authority that exist among indigenous people thus allowing them to work efficiently within these structures and developing trust between all parties involved.

There are seven indigenous groups within the Republic of Panama – the Kuna, Ngöbé-Buglé, Emberá, Wounaan, Teribe, and Bri-Bri – and each group has its own socio-cultural characteristics. This paper will provide more detailed information about the traditional administrative structure and the political administrative structure (national government) in the Emberá/Wounaan *comarca* (land of indigenous populations). The facts presented here represent the situation as it stands today. My purpose is to acquaint readers with indigenous Panamanian people. These groups have struggled for survival as well as for recognition by other societies. My aim is to make future collaboration in Panama more harmonious.

## INDIGENOUS GROUPS IN PANAMA AND THEIR CURRENT STATUS

Many indigenous people are emigrating from their traditional rural lands in search of better social and economic opportunities, although some emigrate in order to study or to seek medical attention. They emigrate either temporarily or permanently to urban centres, including Panama City, and the capitals of the provinces of Chiriquí and Veraguas.[2] Most of these people either eventually return to their traditional lands or maintain close ties with their families. In spite of these constant migrations, the erosion of their native tongues, and the adoption of cultural pat-

*The Ngöbé-Buglé*

Historians named these people *Guaymie Indians*, but today, in their struggle for self-determination, this group wishes to be called Ngöbé-Buglé. These people form the largest Indigenous group in Panama, with 127,410 inhabitants representing 63.6 per cent of the entire Indigenous population. According to the 1990 census, 49.5 per cent of those who are ten years of age are illiterate.

For many years, the Ngöbé-Buglé have fought for a Ngöbé region, as have the Kuna and the Emberá. These demands have been met through the adoption of the Ngöbé-Buglé *comarca*, which was approved by the legislative assembly through Legislation #10, 28 January 1997, ratified on 17 March of that same year by President Ernesto Pérez Balladares.[3] This adoption marked a great chapter in the life of a people struggling for self-determination.

The Ngöbé-Buglé *comarca* has a surface area of 6,944 square kilometres and is located in the west of Panama. The *comarca* lies within the provinces of Bocas del Toro, Chiriquí and Veraguas, and is governed by a general congress and regional congresses. The Cacique general (general chief), the Cacique regional (regional chief), and the president of the general congress are the region's highest authorities. Banana producers, mining companies, large landowners, and hydro-electric companies are now showing great interest in the Ngöbé-Buglé region.

terns that are foreign to their traditions, most indigenous people continue to consider themselves part of their Native group – they do not lose their sense of ethnic belonging. The indigenous people of Panama, as well as other Native people around the world, have struggled for decades to retain their traditional lands and for socio-economic and cultural rights. Their efforts have had a significant impact: for many groups, the recognition of nationhood has been accompanied by the preservation of cultural and linguistic traits. These people have raised their voices before the national government and reached their objectives through a process of negotiation. Today, the larger indigenous groups have jurisdiction over large portions of forested land with great

*The Kuna*

The Kuna are the largest Indigenous population after the Ngöbé-Buglé. These 47,298 people are generally located in the San Blas *comarca*, or the Kuna Yala *comarca*, as they know it. They also live in the Madugandi *comarca*, in the Bayano region of the Province of Panamá, and the Wala-Nurra[4] and Pucuro-Paya regions, both in the province of Darién. The Kuna also live in the Republic of Colombia. The migration of Kuna toward Panama City is quite significant: there are now entire outlying areas populated by Kuna. Kuna women wear their traditional attire at all times in these areas. Over the past fifty years, the Kuna have attained a socio-economic and politico-administrative autonomy within Panama, the result of arduous struggles that culminated in the rebellion of the Kuna in 1925 (the so-called Tulé Revolution). In 1938, the San Blas *comarca* was established, delineated by Legislation #16, 1953, which set forth that the land was both unalienable and untransferable and that the Kuna had the right to manage their own *comarca* – that its organisation and internal affairs should be the responsibility of traditional authorities.

The Kuna Yala *comarca*'s territorial area covers 2,357 square kilometres. It is used for agriculture and livestock. The Kuna communities are located on the Islands of the Mulatas Archipelago. Kuna Yala has a great wealth of flora and fauna, ranging from the summit of the San Blas mountain range to the shores, islands, and coral reefs of the Caribbean Sea, with its rich marine ecosystem.

The *comarca*'s highest administrative authority is the *cacique*, who falls into the same category and has the same powers as the provincial governor. The *sahilas* are the leaders of the community and the keepers of the peace. There is also a general congress, and its highest traditional authorities are the *caciques*.

The Kuna who live in the Bayano region of the Province of Panamá have also obtained official demarcation of their territory, decreed by Legislation #24, 12 January 1996. Approximately 3,500 people live in the Madugandi *comarca*.

ecological wealth. Many groups also control the preservation of their cultures and enjoy relative autonomy from traditional forms of government.

## THE PROVINCE OF DARIÉN

The Emberá people inhabit the province of Darién. This province is located east of the province of Panama, and is the physical link between Central and South America.

Indigenous peoples have always lived in many of the most important areas of natural value, areas that still exist. The Emberá live in one such area – the province of Darién. The National Government's Executive Decree #21 of 1980, created Darién National Park. This park was named a World Heritage site in 1981, and a Biosphere Reserve in 1983 because of its actual and potential biodiversity. It is located in the eastern-most third of the Province of Darién, near the border with the Republic of Colombia and on the limits of Los Katios Natural Park. At 579,000 hectares, it is the largest of Panama's national parks and the largest single-country park in Central America.

Darién is considered "virgin land" by Panamanians, given the vast expanses of forest and the abundance of existing natural and cultural wealth within it. Because it also happens to be the only portion of the Pan-American Highway which to date has not been completed, it is also known as the "Tapón de Darién" (the Darién plug).

## THE EMBERÁ CULTURE

Both traditionally and in scholarly texts, the Emberá were known as "Chocoe" Indians. As part of the group's struggle for self-determination over the past few decades, they have chosen to now be called the Emberá. To be called Chocoe now is considered an insult by the Emberá – the term Chocoe does not even exist in the Emberá language. The literature describes this group as being made up of two well-defined, distinct cultural families of similar linguistic descent – the Emberá and the Wounaan.

According to the 1990 census, the Emberá, with a population of 14,659, and the Wounaan, with a population of 2,605, live throughout the province of Darién. The land they inhabit, which is legally recognized by the national government as the Emberá *comarca*, is located in areas of high community concentration. The Emberá also live in Piriatí and in Ipetí, in the Bayano region, and on the shores of the Río Chagres. Both regions are in the province of Panamá.

During the last decades, they have also been found in the Province of Colón. As is the case with the Kuna, small suburban Emberá establishments are located outside of the country's capital, but the Emberá presence is less obvious in this region because, unlike the Kuna, the Emberá do not always wear their traditional attire. In some Panamanian communities, the Emberá live with the Wounaan. The Emberá also live in the Republic of Colombia.

## SOME EMBERÁ AND WOUNAAN SOCIAL TRAITS

Men of these two tribes traditionally wear the *taparrabo* or *wayuco* (*andi~á* in Emberá) – a small piece of cloth attached by a string worn around the hips – although most have adopted Western dress and wear shirts and trousers. Women traditionally wear the *paruma* (*wa* in Emberá), which consists of a piece of cloth worn around the hips, falling to the knees and exposing the upper body. Traditional dress is worn more often by women than men.

Emberá dwellings are currently clustered into communities (*puru* in Emberá). Each community has its own name. Currently, there are very few dwellings scattered along the shores of the river, which was how the Emberá once lived. The creation of larger communities is a response to government initiatives that seek to provide the Emberá with better health, education, and other services, as well as a result of the influence of missionaries. These dwellings and communities are located on the river's edge, and are always near a mountain stream, for easy access to water.

The houses are usually built on poles made out of such fine wood as balsam, *kira*, or *cocobolo*. The circular roof is held up by four main posts, each measuring 20 to 25 feet in height. Depending on the size of the house, the floor rests on nine poles measuring 7 to 10 feet high. The circular shape of the dwelling can measure between 15 and 20 metres in diameter. The houses have no walls and are completely open. The floor is covered with palm leaves (*eva* in Emberá), and the roof is made of *doquedua*, *guagara*, *potorrico*, or *naguala* – the leaves of forest palm species that are both fine and resistant. All of the material used for building a traditional dwelling is taken from the forest. Today, some Emberá families have adopted Western house-building customs, including adding zinc roofs and wooden or concrete walls and floors. According to data obtained from community inhabitants, securing materials for building traditional dwellings is becoming more and more difficult. The Emberá claim that the species needed grow far from the community and that it takes too much time to gather the materials. We can see through the construction of their dwellings that the Emberá need the forest for many of their activities. Without the forest, the Emberá cannot survive.

The rural and community health programs of the department of health's integrated system insist on immunization, the control of infectious parasites, family planning, courses on nutrition and health, and the organization of committees within each community. In spite of recent efforts, some community health posts do not have the medical provisions needed to offer basic health care services. Very few communities have sanitary water services, for example, since a large majority of people use and consume water from the river or from nearby streams.

Although national literacy campaigns have been carried out in some communities, the illiteracy rate remains high. Since 1975, the ministry of education has been promoting bilingual education programs that give priority to indigenous languages and mother-tongue literacy programs for indigenous teachers. These programs have not been fully implemented, however. Despite the fact that the province of Darién has historically been a disadvantaged region and therefore underpopulated, over the past thirty years it has experienced population growth as a result of the migration of people from the central provinces. These people are called Colonos.

## THE SCIENTIFIC COMMUNITY
## AND THE INDIGENOUS COMMUNITY

Throughout history, indigenous people have existed and inhabited certain territories. Today, they are considered to be the most important custodians of the earth, and hence serve as a point of interest for the scientific community, especially in the fields of anthropology, botany, environmental studies, and geology.

In the past, scientific research was not done in order to inform the average person. In other words, if human beings served as subjects for research, they were not entitled to an opinion about or say in that research. Thankfully, today the subjects of scientific experiments have rights – often encoded in law – that give them a voice in the research to which they are subject.

*An interview with Bonarge Pacheco,* Cacique *of the Emberá Collective Lands of Ipetí, about the relationship between the Emberá and Visiting Researchers*

I am very worried. Years ago, before I became Cacique, eight scientists from the US came here to carry out research. We never knew what they were doing, what they had done. Then, just as I became Cacique in 1992, a woman came here. Her name was Cynthia. She was also from the US and wanted to do research in Ipetí. We decided that we would allow her to come and work here, provided that she made an agreement with us. We wanted her to leave the results of her work with us. This way, we were not denying her the possibility of carrying out her research. Cynthia accepted this agreement. She came to study our lives, that of the Kuna, and that of the colonists living on the other side of the river. Since 1993 we have been waiting for her results. She never came back; she didn't give us anything despite her promise to do so. The only anthropologist who ever complied with her promise is Prof. Esther Urieta of the University of Panama. She came back here one day with Mr. Hugo Giraud, director of Indigenous Politics at the Ministerio de Gobierno y Justicia, to give us a copy of her thesis. So I can tell you that the community has become very distrustful of scientists because we have not seen any benefits from their visits.

In 1992, a group from the University of Panama came here. There were six Americans, two French, and one Chinese researcher. They wanted to study medicinal plants and natural dyes. There was no way we could accept that request. When I refused permission, the professor told me that I was ignorant, and that we had to share because he

Emberá communities have accumulated countless negative experiences – interspersed by some positive ones – with foreign scientists. The Emberá often referred to the scientists as *achir~a* (friend), or *dau paw~ara~bea* (blue-eyed ones). These *gringos* – what Panamanians and the Emberá call any fair-skinned, fair-haired individual – would come to the Emberá communities without any established protocol, did not meet with community officials beforehand, and did not wanted to write a book that could be useful to everybody. Medicinal plants are the only topic on which we refuse to co-operate. If foreign scientists learn our medicine, they will sell it. Then they will make money and we will have nothing.

At the same time, we are aware that this knowledge could disappear. In our community the elders agree to teach the young people. This way we will not forget. It would be great if the leaders in the community could work with indigenous professionals to make a book on medicinal plants. But we have a lack of professionals. We have only one young man studying. He will become a lawyer.

We do not mind research. Research is good and could help people. Even if it is not directly linked with our lives, it is interesting. We could learn, and exchange ideas. The one thing we do not want, though, is for scientists to try and change our way of life. Some churches have done that: they told us that we had to stop wearing our parumas and our jewels. This is dangerous for us – this way we could lose our culture.

The problem with foreign scientists is that they say one thing and do another. For example, a white man came here. He said he just wanted to walk in the forest. So he went with one of us. In fact, that man was studying platanillos (Heliconia). He told us only upon his return. He could have worked on medicinal plants and we would not have known it! So the greatest difficulty with scientists is that they do not tell the truth.[4] They should tell us what they really want to study, and then give us their results. If we had the results, we could make a library and keep the information here so that our youth can learn. Besides, if scientists do not tell us what they study, how can we help them?

explain the purpose of their visit. The scientists asked many questions in order to obtain information and to learn about the culture, but the Emberá simply became passive informants who never questioned the scientists' methods. The Emberá even lugged the researchers' heavy tools and equipment from one site to another. The relationship between the scientists and the Emberá was thus clinical, without alternatives, and practically without value for the Emberá community.

If you ask an elderly person in the Emberá community what he or she remembers of the *gringos* who passed through, they will tell you that the scientists took many pictures, of everything and of everyone in the community, and that they often promised to send them things from far away, but that that promise was never fulfilled. The Emberá enjoy having photographs of themselves and of their families, and so they were keenly disappointed when the scientists failed to send the photographs that had been promised. Other elderly people remember being questioned about their eating habits, the number of children they had, and what their daily activities were. Most of the researchers never offered explanations as to why they wanted to learn so much about the Emberá, although a few did say that they would be writing a book. My grandfather feels that, although the scientists may have tried to explain their intentions, he was not able to grasp what they were saying. The only way to understand what the *achir~a* were doing was to learn what they learned, by attending school. A great *gringo* friend once told him that he should fight for this land, because someday he could end up without anywhere to live. This is one incident that my grandfather remembers vividly. He even recalls this man's name – Harold Becker – which is significant, given how difficult English names are for the Emberá to pronounce.

Over time, new questions began to be asked by the Emberá: why do the wealthy *achir~a* travel great distances to come and visit us, the impoverished Emberá? What do they do with the things they take from us? Why don't they take our children back to their countries and educate them? Gradually, distrust began to build, as Native people began to understand the one-sidedness of these relationships.

In a more equitable exchange, all aspects of the project would be taken into account, from formulating the study to the publishing of results. For example, in order to undertake research, consultation with and consent from local people must clearly be obtained. The local people should also be told what is to be done with the research results and what benefits they might reap from such research. They should also be given a written account of the results.

Today, not only is a permit required to visit indigenous people in Panama, but the researcher must also obtain a permit to visit the territories of the conservation parks, places where no humans live. In this way, an equitable exchange is clearly being cultivated.

## REACTIONS OF INDIGENOUS PEOPLE
## TO RESEARCHERS

It is common knowledge, especially within the scientific community, that indigenous people have been living for centuries on land
that has been passed on to them by their ancestors, in many
important natural areas that understandably provoke a great deal
of interest. The fear that permeates the indigenous community is
grounded in the belief that although they have co-operated with
researchers for many years, the value of their knowledge has
never been acknowledged, and today, more than ever, they wish
to preserve the last legacies of their traditional culture. The
indigenous people therefore have a dislike of researchers clearly
summarized in the following passage: "These people [the scien-
tists] come to our land and do anything they please; they learn
from us, teach us nothing in return, and then we never hear from
them again."

For many years researchers have come to carry out research on
the Emberá; sometimes the people knew what the research was
about, and sometimes they did not.[5] To this day, very few study
results have ever been sent back to the Emberá by researchers who
worked in the community. Although this literature can be found in
libraries and bookstores, such resources are not accessible for most
indigenous people.

## THE SUSPENSION OF SCIENTIFIC
## RESEARCH BY THE EMBERÁ GENERAL
## CONGRESS

Law #22, 1983, led to the creation of the Emberá *comarca*.[6] In one
of its clauses, this legislation established the *comarca*'s General
Congress as its highest traditional decision-making body, which
speaks on the Emberá's behalf. Regional and local congresses – traditional bodies of expression and decision – were also established.
The council of *nokoes* (community leaders) was established as well,
and serves as a consulting agency for both congress and *comarca*
leaders.

The general congress, which was held in April 1995, occurred in
response to outside requests for the implementation of projects with
a scientific research component in the indigenous areas within the

*comarca*. The congress suspended all scientific research in the region's area until further notice. Similar restrictions were adopted by all communities outside of the *comarca*'s jurisdiction. In other words, no one can go to an indigenous community without contacting the proper authorities. This measure is a result of the negative experiences that indigenous people have had with research communities over the years. Today, indigenous people demand respect for their traditional knowledge and the preservation of their cultural values.

This fear – expressed in a concern for ancestral know-how and a claim to intellectual rights – is prominent. Given the current will to understand and respect indigenous peoples' rights around the world, Native peoples may need to learn how to conquer that fear and work with scientific communities, government bureaucracies, and private institutions. This will open up possibilities for negotiation and a new kind of dialogue. Active participation by indigenous people in all scientific research can also become the basis for mutual agreement and trust between the scientific community and the indigenous community.

### IMPROVING THE RELATIONSHIP BETWEEN THE SCIENTIFIC COMMUNITY AND THE INDIGENOUS COMMUNITY

Every scientist has a distinct culture, is a citizen of a particular country, and views the world – and research subjects – through a unique political lens. Based on this understanding, I have developed guidelines to help improve the relationship between scientific and indigenous communities.

### GUIDELINES FOR RESEARCHERS WORKING WITH INDIGENOUS COMMUNITIES

- Scientists must understand the levels of authority present in an indigenous territory before developing study plans and beginning work. In Emberá communities, there is a *cacique* within the *comarca* who is the foremost traditional authority for the Emberá people. His functions and powers are set forth in the *comarca*'s constitutional law. The *cacique* is the principal representative and spokesperson for the *comarca* before all non-indigenous groups, and therefore must be the first person that researchers consult in the community. In chapter two, Marie-Hélène Parizeau underlines the need to obtain consent from Native people when working within their areas.

- The lack of professional ethics that researchers have shown when working with indigenous people should be recognized as a form of injustice that constitutes a historical indebtedness. A firm determination to redress this situation, particularly the withholding of information and research results from indigenous groups, needs to be assured.
- All research initiatives must be based on deep respect for indigenous peoples, their diversity, and their wealth of knowledge and tradition.
- Future research and projects with indigenous communities must be based on respect, the full participation and knowledge of the Native peoples, and shared responsibility.
- The indigenous peoples must have the right to participate on an administrative level in any research or scientific project and to hold intellectual property (copyright) and patent over study results.
- All draft statements of indigenous rights drawn up by international agencies and Western governments should be made specific.
- Universities or other educational institutions must provide study opportunities for indigenous people so as to balance out their participation in research projects.
- Before launching any activity in an indigenous area, interested parties must first submit the project's goals and must come to an agreement with the proper agencies and authorities in the intended work area. The execution of an agreement and the levels of authority to be consulted will depend on the scope of the project or study proposed.
- Upon completion of a project or study, copies of all reports must be submitted to the indigenous community and to the proper authorities. Such documents should be educational and informative for the entire community.

NOTES

1   *Censo Nacional de Población y Vivienda* (Panamá: Contraloria General de la República de Panamá, 1990).

2   *Censo Nacional de Población y Vivienda* (Panamá: Contraloria General de la República de Panamá, 1990).

3   Dirección Nacional de Política Indigenista, Ministerio de Gobierno y Justicia, República de Panamá.

4   By the time the book went to press it had become the Wargandi *comarca*.

5   An example of this is James A. Duke's method of extracting ethnobotanical knowledge from the Kuna of Panama: "Using the Cuna (Kuna) concept of mankind to psychological advantage greatly facilitated the author's inquiry . . . Cuna dislike Choco (Chocoe) because they believe them to be naked devils (nia) who kill more game than they need. By playing on this enmity, you can learn many Cuna ethnobotanical secrets. Do not ask a Cuna how he uses a plant. Instead, tell him how a Choco uses it. The Cuna will then play down the Choco usage, and tell you the Cuna usage. Information is thus volunteered by, not extracted from, the Indian, and the information is probably reliable." J.A. Duke, "Ethnobotanical Observations on the Cuna Indians," in Economic Botany, 1975, 29: 278–93.

6   Law #22, Republic of Panama, 8 November 1983.

## 4 Cultural Lenses and Conservation Biology *Collaboration in Tropical Countries*

PRISCILLA WEEKS, JANE PACKARD,
AND MIRELLA MARTINEZ-VELARDE

### EDITORS' INTRODUCTION

The need for scientists to fully explain their work to the communities they are studying was the primary message of chapter 3. *A priori*, this task seems simple, yet it requires a mutual understanding of key terms, concepts, and values. In this chapter, Priscilla Weeks and her co-authors examine the importance of cultural perception to the determination of the value assigned to biological diversity. Scientists working in foreign countries must understand that their local colleagues and the local population may view study objects quite differently. A failure to recognize this can lead to a discouraging and counterproductive *quid pro quo*.

### INTRODUCTION

This chapter focuses on the communicative aspects of field work in conservation biology. Conservation biology is both a social and a scientific enterprise – fieldwork abroad brings biologists into contact with an array of scientific and nonscientific social groups. Not all researchers are equally successful in negotiating these relationships: while some recognize the social nature of conservation biology, many who focus on scientific excellence may feel ill-equipped to consider how scientific collaboration both affects and is affected by social issues.

Two guiding concepts are used in this chapter: cultural lenses and reflectivity. A cultural lens refers to the common knowledge, general cognitive frameworks, and values that are (more or less) shared by the members of a social group and that help them make sense of their world. In the context of international field work, a social group may consist of villagers, scientists, government employees, funders, or politicians. In the first part of the chapter, we discuss the nature of cultural lenses and how they affect a scientist's work. We then discuss reflectivity – now biologists must be aware of and critically analyze the lenses that guide their conservation efforts. Communication occurs within the context of social relationships, and biologists who are trained to be distant from their research subjects must learn how to become engaged in reciprocal and mutually beneficial collaborative projects.

Although scientists must deal with many people, from politicians to peasants, this chapter focuses only on potential collaborators. There are two broad groups with whom scientists usually collaborate in cross-cultural field work: other scientists (university, government, or non-governmental organizations (NGOs)); and local communities. Neither of these groups is homogeneous: there are structural and ideational factions within each group that affect how its members view and interact with foreign scientists.

The Panamanian scientific community's relationship with the Smithsonian Tropical Research Institute (STRI), an American research institute based in Panama, is a good example of the problems that may arise between foreign and local scientists. STRI is one of the world's leading basic research centres on new- and old-world tropical organisms. Its international staff of over thirty scientists conducts research into animal behaviour, plant ecology, canopy biology, conservation science, and genetics. STRI traces its history to the construction of the Panama Canal and the control of insect-borne diseases such as yellow fever and malaria. Scientists at STRI studied the flora and fauna implicated in the spread of these diseases and then became interested in establishing a permanent biological reserve on Barro Colorado Island.

The scientific community in Panama is divided over the value of hosting the Institute. In informal communication, the community discusses the degree to which STRI is elitist: most resident staff scientists are expatriates working on basic research in molecular and evolutionary biology. As "pure" scientists, they do not actively seek solutions to problems that are either relevant to local communities or identified by Panamanian scientists as important to the country. "Applied" Panamanian scientists feel that this emphasis on basic biology devalues the type of applied biology that interests them and that directly benefits their country.

From another perspective, however, the biological sciences are the most developed branch of science in the country precisely because of STRI's significant contribution to the field.[1] Additionally, many individual Panamanian scientists have directly benefited from the Institute. Local scientists have access to its library and computer facilities and STRI has provided a networking avenue between local and international colleagues. Finally, some STRI staff scientists have developed applied research programs that meet Panama's conservation needs. It is clear, therefore, that Panamanian and foreign scientists are not simply two distinct, monolithic social groups.

The complexity illustrated by the above example is true for all of the examples cited in this chapter. Cultural lenses may vary among groups within a country, as well as among countries. The intricacies of intra-group dynamics for each case are not examined in this brief discussion, although a great variety of relationships resides within the cases studied.

### CULTURAL LENSES

We chose the word lens (instead of perception or mental model) for its metaphorical value. A lens both focuses and filters: it focuses on important objects and filters out extraneous ones. Forest-dependent villagers, for example, may look at a forest and see fuel, medicine, fruit, housing materials, shade, fodder, and, perhaps, a sacred space. An ecotourist may see a place of adventure, an area in which to view animals, or an unspoiled place to "commune with nature." A lumber company and its employees may see the forest as a source of income. Cultural lenses are thus acquired through formal education and professional socialization, as well as through life experience.[2]

Scientists acquire lenses from their educational experiences and professional interactions that teach them how to frame appropriate scientific questions. Through general scientific training, scientists learn how to identify problems, decide the kind of information needed to solve a problem, and gather and analyze this information. Each discipline also imparts a set of values to its members, a language (or jargon), a knowledge base, and a set of correct practices.

For example, evolutionary biologists, community ecologists, wildlife biologists, and foresters acquire different sets of lenses that focus their attention on particular questions. An ecologist sees a forest, for example, as a particular type of ecosystem. To a geneticist, however, a forest is a gene pool. To a wildlife biologist, it is a habitat. A forester views the forest it in extractive terms and in terms of tree health. Each discipline thus attempts to "pay attention to different facts and make different sense of the facts they notice."[3] As well, different knowledge bases are based on different values. A forester not only knows how to grow trees, but also values the greater production, superior lumber, and ease of harvest provided in a monoculture of hybrid pine trees. An ecologist not only understands ecosystem functions, but also values the biological diversity that distinguishes an ecosystem from a monoculture. An ecologist probably would not consider a monoculture a forest, but a forester likely would.

Conservation biologists also view the world through a unique set of lenses. These lenses can be understood through textbook review and the professional society's statement of goals. Such documents express a scientific social group's ideology most clearly because they are used to express existing members' values and knowledge and to socialize new members. For example, introductory texts portray conservation biology as a multi-disciplinary science that includes a subset of biological specialties and is informed by the social sciences, law, and humanities.[4] Although it is a young discipline that is still defining its core values, it is guided by these "normative postulates": 1) the intrinsic worth of biological diversity; 2) the value of ecological complexity; 3) the "goodness" of evolution; and 4) the "wrongness" of anthropogenically caused extinction of species.[5] These assumptions are encoded in the Society for Conservation Biology's statement of goals: to "help develop the scientific and technical means for the protection, maintenance, and restoration of life on this planet, its species, its ecological and evolutionary processes, and its particular and total environment."[6] The value judgments intrinsic to the postulates of conservation biology, as well as their ethical implications, are examined by Marie-Hélène Parizeau in chapter 2.

*Local Involvement in Conservation Decisions in Mexico: an example*

In a course designed to increase professionalism and networking in the management of protected areas in northern Mexico, the issue of local involvement in conservation was actively debated. Julio Carrera, a co-facilitator of the course, encouraged participants to think about how different a protected area might be if the impetus for its establishment came from the desire of local residents to protect their environment rather than from a distant government office. To illustrate his point, Carrera used the example of Boquillas, a small Mexican settlement across the Rio Grande from Big Bend National Park. Boquillas lies within a protected area that was established as a "sister" park to Big Bend. International sister parks were the vision of a park superintendent who was actively involved with the planning of Waterton Lakes in Canada and Glacier National Parks in the United States.[7] Villagers of Boquillas subsist on the wax harvested from a particular plant in their area, and Carrera was consulted on how the harvest could be made sustainable. The methods he suggested were not implemented, however, since, for a villager leading his donkey in the hot sun, it was impractical to pass by one clump of the plant in order to forage on a more distant patch with the hope that the first clump might be there in the future.[8] In this example, practical conservation options could only be found through the active participation of local people in the consultation process.

Through personal exchanges and philanthropic programs, the staff at Big Bend and at a Mexican non-governmental organization (NGO) have cultivated an understanding of how villagers make a living and have talked about conservation and the future of natural resources with local people. Although progress in local conservation efforts may be slow, it is hoped that it will prove to be more enduring in the long-term.

Conservation biology's assumptions separate it from the traditional, conservation-related disciplines of forestry, wildlife, and fisheries management. The belief in nature's intrinsic value, in particular, is not a key tenet of the management sciences (chapters 5 and 6 discuss the consequences of this belief). Conservation biologists believe in long-term preservation while management scientists focus on sustained yield; conservation biologists pursue the health of ecosystems, while management scientists take a species-by-species approach. Thus, conservation biology's norms may filter how its scientists interpret a collaborator's statements.

Shared systems of meaning and cultural lenses can thus aid communication within a group, but impede communication between groups. For example, the belief systems of pre-Conquest cultures in Mexico included elements of both reverence for and utilization of nature.[9] A conservation biologist may, during dialogue with a Native, filter out the utilitarian aspects and focus instead on the similarities between reverence and intrinsic value. A wildlife manager, however, may filter out reverence and focus on utilitarian harvest. The conservation biologist and wildlife manager may thus communicate better with the Native person than with each other. The sections that follow illustrate this fundamental impediment in greater detail.

## REFLECTIVITY BETWEEN COLLABORATORS

Steier[10] used the term reflexivity to explain how communication turns back upon itself – how it is both shaped by and shapes the relationship in which it is expressed. He emphasized the importance of researcher self-awareness when studying communication processes, and concluded that reflexivity "allows us to open outward toward other forms of this self expression, and toward an awareness of the multiple conversations in which we and our reciprocator participate." Such awareness can be beneficial for both practitioners of social change and those who study the processes of that change.[11] Conservation biologists advocate change in social processes that disrupt what they see as natural, ecological, or evolutionary processes. Thus, Clark, Reading, and Clark[12] urged conservation biologists to become reflexive practitioners.

As with our choice of the word lens, we chose the word reflectivity for its metaphorical value: the light that a lens both filters and reflects is analogous to the ways in which we give information meaning through the understanding of patterns. To reflect is to cast back a mirror image: clouds mirrored in a lake are not those clouds, yet their patterns of light are very similar on the retina. By analogy, if I sense that my collaborator shares certain values, I can communicate through those values and make it easier to understand the cultural lens of my collaborator and for us to focus on the problem at hand.

Reflection also involves careful consideration. For example, when two mirrors face each other on either side of the study object, reciprocal images can be extraordinary. Careful consideration is needed to decide how this reflected information will affect behaviour and to trust that a collaborator will respond likewise. Using this information to change a collaborator's behaviour is not considerate, and will likely be damaging to the relationship.

The context in which conservation biology is practised is not universal: scientists from Southern and Northern countries face different rewards and constraints, even when working on the same project. Both foreign and host scientists are responsible to, and must balance the demands of, multiple constituencies (see chapters 1 and 3, for example), including local communities, funding agencies, and collaborators, as well as competing interests within universities and government departments. Furthermore, local and foreign collaborators may not hold similar positions in their respective institutions: in countries with few Ph.D.-level scientists, for example, Masters-level scientists are matched with Ph.D.-level foreign scientists. Extension-oriented applied scientists may be matched with scientists who are primarily researchers and who are interested in answering basic questions. These differences often affect the prestige and power of research team members: local scientists who should be equal partners in a project may be treated like research assistants instead of scientific collaborators. (This point was vividly discussed by Léonard Mubalama in chapter 1.) These pressures can lead to miscommunication and frustration between research team collaborators, and the failure of projects.

As well, unrealistic expectations may arise from the failure to understand the different scientific and social contexts in which conservation biology is practised.

## SCIENTIFIC INFRASTRUCTURE
## AND RECIPROCITY

Scientists in Southern countries often use international collaborative projects to strengthen their countries' existing scientific infrastructures or resources, both human and technical. Collaborative projects can aid in improving a library or a laboratory, provide access to foreign funding, and introduce students to new ideas and new techniques.[13] Collaborative projects can also open up access to primary information from abroad, an important consideration for institutions with limited funds. Sometimes local researchers' interest in hosting collaborators arises as much from these interests as it does from interest in a particular research project, especially when a project does not address the priorities of the host country.[14]

The primary goal of collaboration, therefore, may be to develop research capabilities in a certain area as a first step away from dependence on foreign researchers. Foreign scientists who thus focus only on the project at hand fail to fulfill these expectations. Filipino scientists often described this kind of scientist as a "hit-and-run" researcher: someone so involved in building up his or her career that he or she views the research project as private property and does not recognize other scientists' interests and/or roles in that research. Data and samples are taken back to the visitor's institution to be analyzed, for example (see chapter 1), reprints are not sent back to communities in the host country, and colleagues in the host country are not acknowledged in, or treated as co-authors to, publications. In chapter 3, Rogelio Cansarí suggests that this kind of negligence is largely responsible for the dislike his people have developed toward foreign researchers. The development of research projects is often just one of many responsibilities for scientists in developing countries. Rearranging busy schedules to make arrangements for visitors or accompanying them to the field can represent a considerable investment for host scientists, and failure to reciprocate is seen as lacking collegiality.

Failure to treat the host scientist as an equal member of a research team is also not acceptable, although foreign scientists may cite many reasons for their failure to cultivate reciprocal relationships. For example, a key cultural filter for a research scientist is the inherent value of building a knowledge base. In the scientists' view, sharing research results through peer-reviewed journals fulfills the scientist's responsibility to peers (and, we might argue, to society itself). Furthermore, a scientist who wants to continue doing research and publishing articles faces a very real pressure to maintain levels of funding, to pursue "hot" research topics, and to train graduate students (all these pressures take on a sense of urgency if the scientist does not have tenure or faces post-tenure review). These responsibilities to one's university and career may, therefore, curtail a scientist's ability to invest time in activities that are not directly related to his or her own research and publishing, such as writing reports for the host country in the national language, sponsoring host country scientists for further training, giving seminars, conferring with scientific leaders about the project, and designing projects that host-country scientists view as relevant.[15]

A second area of miscommunication stems from a lack of understanding of a collaborator's position in his or her respective academic or conservation community. For example, Packard obtained seed funding to accompany a USAID student to his native country of Cameroon. The student was on faculty development leave, and the purpose of the visit was to explore potential collaborative research projects that could be initiated when he returned. Packard naïvely talked to administrators in the host institution about developing a "memorandum of agreement" between their respective universities. The project was on track until one of the host administrators asked about the amount of money that the institution would be receiving for this collaboration. Packard explained that the agreement would be to write proposals to ask for funding from conservation organizations, not to administer funds already obtained. That faculty members might seek their own funding was not in the realm of experience of the administrators, who had previously been supported by a development project. With empty pockets, Packard's prestige dropped, and the doors to the project were quickly closed.

The example of Cameroon, and other complaints voiced about hit-and-run researchers, highlight how important it is for potential collaborators to be frank about their expectations and the constraints they face in meeting those expectations. Administrators in a host institution may be outside the focal distance of the lens of a visiting researcher, so in international fieldwork the astute researcher would be advised to use several lenses.

FRAMING ISSUES THROUGH
MULTIPLE LENSES

The social context in which scientists practise also affects how they frame issues and solve problems. "Framed" refers to the manner in which parts are put together in a whole. Socialization into scientific culture does not replace national culture: lenses acquired through scientific training coexist with those acquired through lived experience, and the latter lenses affect the way in which the scientific lens is focused. Jennings[16] describes how agronomists made decisions during the green revolution about corn and wheat production in Mexico, decisions that precipitated major social consequences. These American scientists focused on the selection of high-yield varieties of corn in order to ensure an adequate supply of food in Mexico. Because of their agronomic background and training, they identified the task in terms of higher productivity: specific tasks, for example, included weed control and the development of hybrid varieties. In doing so, however, they defined the problem as one that could be solved strictly within their area of expertise: the social, political, and economic contexts in which higher production take place and that affect food distribution and the livelihoods of less well-off farmers were not within the agronomists' frame of problem solving. The cultural lens that focused their attention on higher productivity filtered out non-agronomic information, which they considered to be ancillary to their goals.

If the question of food resources in Mexico had been framed differently, with an understanding of the local social context, a different solution might have been found. Indeed, some of the Mexican scientists involved in the project did see the question of higher production in quite different terms. They favoured, for example, strengthening local varieties of corn over the production of hybrids, and the promotion of subsistence crops over commercial crops. According to Jennings, these scientists favoured a research agenda that could provide a production strategy for *campesinos* (small-scale farmers). In contrast, foreign scientists framed the issue in technical terms, allowing them to escape responsibility for the social processes triggered by the improved production system they designed – namely, the strengthening of "elite" farms and farm production methods. "While local scientists argued the merit of agrarian reform, Rockefeller Foundation supported scientists limited their activity to experimental plots."[17]

Conservation biologists who focus only on the biological aspects of conservation do the same type of filtering, and run the risk of inadvertently privileging one aspect of a social system over another. One of conservation biology's key premises – the intrinsic value of nature and biodiversity – is especially problematic for local communities[18] (see chapter 5). For example, a common tool used to save biodiversity in a particular area is the institution of protective status. The action of preservation can, however, favour tourists over resource extractors, settled agriculture over swidden, and corralled livestock over grazing animals. Priscilla Weeks was told by Indian villagers who had been displaced by a Project Tiger reserve (a national government program with significant American backing) that Indian forests belonged to foreign tourists. The decision to remove villagers from the forest had both social and conservation consequences, however. First, villagers were moved to degraded areas and consequently suffered economic as well as physical displacement. Second, after the move they felt they had no reason to help protect the forest, and so frequently engaged in the illegal harvest of grasses and fuelwood.

The example of Boquillas (see p. 45) illustrates the link between the social system in which the scientist is embedded and the conservation approach that is favoured. In Southern countries, recognition of the social system also implies recognition of the needs of local communities. The displacement of residents, which occurred, for example, in parks that were established in the United States at the turn of the century, has been critiqued as being inappropriate for sites in Southern countries, because communities live in and around these areas. This communal presence can actually change the way in which scientific information is used. Even in the United States, biologists would be wise to understand the way in which local communities perceive nature and how they depend on it, especially when these factors are at odds with scientific understanding of appropriate conservation action.[19]

### LOCAL COMMUNITIES AND REFLECTIVE FIELDWORK

Conservation biology affects local communities both directly and indirectly. Biologists who are active in the design of protected areas, the articulation of indigenous taxonomies, or the establishment of co-operative management work directly with local communities at different levels of involvement and commitment. Taxonomists, for example, may rely on the expertise of knowledgeable persons in a community, may view their involvement with a community as limited, and feel that they have adequately repaid a community through payment for services or through bringing local resources to the attention of policy-makers, as a first step to saving these resources. This limited view of reciprocity, however, may not be shared by the community, since, as we have seen, scientists and local communities usually differ in their definition of a research project's desired outcome: scientists value answering basic scientific questions, while the community may need answers to pressing problems (such as a shortage of local health-care facilities). Faced with problems of land tenure, water shortages, and degraded lands, for example, local peoples may have difficulty understanding the need for basic research (such as taxonomy) and the restricted roles, vis-à-vis community problems, that scientists have constructed for themselves.

Local peoples also often see scientists as potential resources to help them solve "on-the-ground" problems. Foreign scientists may have access to aid agencies and government officials, and so local communities may mistakenly believe that a visit by these scientists signals an aid agency's interest in the community. In other words, basic scientists are often viewed as community developers. This misunderstanding may result in the placing of unrealistic expectations on foreign researchers, and misperceptions of their role may lead to negative views of scientists who are unwilling or unable to fulfill community expectations.

Biologists who work in protected areas where locals reside, or who are looking for alternatives to certain modes of extraction, face a different set of community issues. These scientists may accept the need for a long-term commitment to the local community, but may not have the training or inclination to successfully involve community members in conservation or development projects. Although it is outside the scope of this chapter to fully discuss participatory ecodevelopment, see Wigley and Baser, chapter 7, and Butler and Regis, chapter 8 for a fuller exploration. Although an array of methods for working with communities exists,[20] scientists must be wary of social engineering that privileges the views of those who work outside of local communities (i.e., the visiting scientist). Truly participatory processes allow community members to set their own goals and priorities, not just design the processes by which others' (i.e., the conservationist's) goals are met.

The previous paragraphs have looked at problems that arise when biologists work directly with communities. Even conservation biologists involved in more "basic" research, such as scientific taxonomy, population biology, and genetics, can affect local communities, however, and therefore have responsibilities toward these communities. These more basic sciences frame questions in a way that excludes social issues and, in so doing, scientists often fail to understand the potential effect of their work on local communities. A tiger census, for example, may seem to be of little relevance to a local community. Using this information to design park boundaries, however, may have profound consequences for local peoples. Similarly, the discovery of a particular plant species' biochemical properties can lead to a discovery of medicinal or technological value, which in turn can lead to corporate interest in an area. Such a discovery can also affect the local communities that rely on these plant species. For example, Indian villagers were aware of the recent patenting of genetic material from the neem tree, and became concerned that their access to neem would be restricted. They also felt that one of their resources had been appropriated by interests more powerful than theirs. The first step in both the process of patenting of neem and the drawing of park boundaries, therefore, was the process of discovery arising out of basic research.

According to scientific logic, the scientific responsibility to knowledge production supersedes more on-the-ground responsibility to local communities,[21] which leads to a failure to understand the consequences of knowledge production.

Conservation biologists have many responsibilities toward local communities. The first responsibility should be to structure scientific research and interaction with a community so that disruption to that community is minimal. This can be accomplished through

1 understanding community dynamics, needs, and structures, so as to not cause friction among members; and,
2 tracing out the potential consequences of one's actions, so as to avoid harmful effects to the community. For example, is a decision not to publish the tiger census ever justified if it might lead to the relocation of a village against villagers' wishes? This question is one that can only be answered by the scientists involved, and although this approach seems antithetical to scientific interest, fairness to the community that is at risk of being moved necessitates that the issue at least be confronted.

Anthropologists have had to confront the issue of (what some might consider to be) the suppression of data since the discipline's inception. Despite anthropology's image as a liberating science, knowledge about communities has been routinely used by colonial administrators and intelligence agencies.[22] The social groups studied consider other types of knowledge to be their cultural property, property that is not to be shared with the outside world. Some anthropologists faced with such situations have responded by not publishing all the information they gather on a community.[23]

Conservation biology faces similar ethical dilemmas. Information about species decline, for example, has real consequences for the human communities who share the same physical space as a species in question. These points of tension – between scientific values and human rights – need to be confronted by conservation biologists, who must work with local communities to find solutions.

Other, less controversial, responsibilities to communities include sharing research results in a form usable by the community, maintaining a local collection of ethnobiological materials,[24] helping the community solve problems it identifies as important to it, and helping the community form links with outside interests.

Thus, the cultural lens that scientists bring to conservation research is only one of several lenses needed in any pluralistic society, since local communities may vary in their understanding of the potential implications of scientific knowledge. As well, with greater awareness it becomes more difficult for scientists to maintain the privileged status of detachment. Although reflectivity in collaboration with local communities may be slow and may affect what information is published, working relationships will likely be sustained if long-term conservation objectives are addressed.

CONCLUSION

We have identified a few of the assumptions in conservation biology that are likely to make collaboration difficult during fieldwork. Some of these are general to science – for example, the assumed superiority of the scientific method in generating knowledge about the natural world. Others are more or less unique to conservation biology and its subset of disciplines – for example, the intrinsic value of biological diversity. The role that national culture plays in focusing the lenses of individuals was also explored, since cultural lenses are intrinsic to particular social contexts. International field-work makes it necessary for scientists to work in unfamiliar social contexts and with people who have different points of view. Scientists will be more successful in their collaborations if they recognize and understand these differences. To accomplish this, they must have the ability to analyze the assumptions embedded in their own scientific practice – that is, they must be reflective.[25] A reflective scientist acknowledges that the cultural lens acquired through training in a particular discipline and in a particular cultural context is only one of many possible lenses and that other, equally valid, points of view and problem-solving methods exist. The ability to place one's own assumptions and values within a network of other possible assumptions and values leads to more open communication with members of other social groups. Understanding and valuing the needs and expectations of collaborators is the first step towards forming realistic collaborative expectations and negotiating outcomes that will be beneficial to all parties.

NOTES

1   O. Sousa, "El reto de la investigación científica y la innovación en la Universidad de Panamá," in S. Sanchez and J. Bosco Bernal (eds.), *Desarollo Científico y Tecnológico: Desafío para Panamá* (Rep. of Panamá: Litho-Impresara Panamá, 1996), 17–25.

2   D. Schon, *The Reflective Practitioner* (New York: Basic Books, 1983).

3   D. Schon, *Educating the Reflective Practitioner* (San Francisco: Jossey-Bass Publishers, 1987), 5.

4   G. Meffe and C. R. Carroll, *Principles of Conservation Biology* (Sunderland, Mass.: Sinauer Associates Inc., 1994); R. Primack, *A Primer of Conservation Biology* (Sunderland, Mass: Sinauer Associates Inc., 1995); and M. Hunter, *Fundamentals of Conservation Biology* (Cambridge, Mass.: Blackwell Science Inc., 1996).

5   M. Soulé, "What Is Conservation Biology?" *Bioscience* 35 (1985): 727–34, 730.

6   Society for Conservation Biology, "Goals and Objectives of the Society for Conservation Biology," *Conservation Biology* 10 (1996): no p.

7   A. Chase, *Playing God in Yellowstone: The Destruction of America's First National Park* (New York: Harcourt, Brace, Jovanovich Publishers. 1987).

8  J. Carrera, personal communication.

9  L. Simonian, *Defending the Land of the Jaguar: A History of Conservation in Mexico* (Austin, Texas: University of Texas Press, 1995) 11.

10  F. Steier, "Reflexivity, Interpersonal Communication, and Interpersonal
    Communication Research," in W. Leeds-Hurwitz (ed.), *Social
    Approaches to Communication* (New York: Guilford Press, 1995),
    63–87.

11  D. Schon, *The Reflective Practitioner*; and *Educating the Reflective
    Practitioner*.

12  T. Clark, R. Reading, and A. Clarke, "Introduction," in T. Clark,
    R. Reading, and A. Clarke (eds.), *Endangered Species Recovery: Finding
    the Lessons, Improving the Process* (Washington, D.C.: Island Press,
    1994), 3–18.

13  J.M. Packard and D.J. Schmidly, "Graduate Training Integrating
    Conservation and Sustainable Development: A Role for Mammalogists
    at North American Universities," in M.A. Mares and D.J. Schmidly,
    eds., *Latin American Mammalogy: Topics in History, Biodiversity,
    and Conservation* (Norman, Oklahoma: University of Oklahoma Press,
    1991), 392–415.

14  J. Gaillard, *Scientists in the Third World* (Lexington, Kentucky: The
    University Press of Kentucky, 1991).

15  R. N. Adams, "Responsibilities of the Foreign Scholar to the Local
    Scholarly Community," *Current Anthropology* 12 (1971): 335–9.

16  B. Jennings, *Foundations of International Agricultural Research: Science
    and Politics in Mexican Agriculture* (Boulder, Colorado: Westview Press,
    1988).

17  Ibid.

18  B. Brower, "Crisis and Conservation in Sagarmatha National Park,
    Nepal," *Society and Natural Resources* 4 (1991): 151–63.

19  P. Weeks and J. Packard, "Acceptance of Scientific Management by
    Natural Resource Dependent Communities," *Conservation Biology* 11,
    no. 1 (1997): 236–45.

20  See Michael Cernea (ed.) *Putting People First: Sociological Variables in
    Rural Development* (Washington, D.C.: Oxford University Press, 1991);
    Robert Chambers, *Rural Development: Putting the Last First* (London:
    Longman, 1983); and Peter Poole, *Indigenous Peoples, Mapping and Bio-
    diversity Conservation* (Washington, D.C.: World Wildlife Fund Biodiver-
    sity Support Program, 1995).

21  W.B. Pearce, *Communication and the Human Condition* (Carbondale,
    Illinois: Southern Illinois University Press, 1989).

22  Talal Asad (ed.), *Anthropology and the Colonial Encounter* (New York:
    Humanities Press, 1973).

23  Bea Medicine, "Learning to Be an Anthropologist and Remaining 'Native'," in E.M. Eddy and W.L. Partridge (eds.), *Applied Anthropology in America* (New York: Columbia University Press, 1978), 182–96.

24  A. Gupta, *Dilemma in Conservation of Biodiversity: Ethical, Equity and Moral Issues. A Review*, mimeograph, no d.

25  See T. Clark, R. Reading, and A. Clarke, "Introduction," in T. Clark, R. Reading, and A. Clarke (eds.), *Endangered Species Recovery: Finding the Lessons, Improving the Process*, 3–18; D. Schon, *The Reflective Practitioner*; and *Educating the Reflective Practitioner*.

## 5 Conservation in Action
### *Assessing the Behaviour of National and International Researchers Working in Madagascar*

LALA H. RAKOTOVAO,
R. RAKOTOARISEHENO, AND
CHANTALE ANDRIANARIVO

EDITORS' INTRODUCTION

Madagascar has attracted the interest of European and American biologists for many years because of the bounty and diversity of its flora and fauna. The island was recently classified by the World Conservation Monitoring Centre as one of the world's eighteen biologically diverse "hot spots," and it is the focus of many large-scale conservation initiatives. The long, rich history of contact between foreign scientists and Malagasy people, and current concern for the survival of Madagascar's unique wildlife, make this country an interesting place in which to examine international scientific collaborations in biodiversity conservation. In this chapter, L.H. Rakotovao, director of the Centre de recherche en environnement of Madagascar, and two of her colleagues, R. Rakotoariseheno and C. Andrianarivo, highlight the problems inherent in the implementation of biodiversity conservation policies in a developing country such as Madagascar. They emphasize that the protection of biodiversity cannot be divorced from a concern for the well-being of local populations, and that ethical problems occur when both national and international researchers implement conservation policies.

## INTRODUCTION

This chapter is essentially an assessment of the behaviour of national and foreign researchers in the course of their conservation activities in Madagascar. It discusses both what is expected of conservation specialists, as scientists and administrators of natural resources in tropical countries, and the behaviour of foreign researchers working in developing countries such as Madagascar. It questions whether the behaviour of researchers conforms, overall, to what the Malagasy communities expect.

## THREATS TO NATURAL RESOURCES

In Madagascar, species and ecosystems are threatened mainly by human activities, particularly those performed within the framework of sectorial policies (for example, road construction, agriculture, and tourism), and by socio-economic factors – population growth beyond the sustainable capacity of a region or of a given population centre, population shifts, or uncontrolled industrialization (Madagascar National Biodiversity Report). Endemic poverty also contributes to diminishing biological diversity, since the irrational exploitation of resources, particularly of wood, is common in poor communities.

## CONSERVATION MEASURES

Faced with threats to the natural environment, Madagascar has adopted various conservation measures. The country was once covered by forests, but these have been greatly diminished and now make up less than 25 per cent of Madagascar's land: a single virgin forest on the island's eastern slope remains. The degradation is a result of logging, overdevelopment, and land clearing, and new regulations have been established to protect the remaining forest.

In 1984, Madagascar adopted a nature-conservation strategy. In 1987, this strategy became an Environmental Action Plan, the first phase of which has just been completed. Quite recently, the island has adopted a Charter for the Environment. Conservation measures on site preservation have been developed through the establishment of protected areas, including the Mananara-North Biosphere Reserve and thirty-nine other sites.

*Biological Diversity in Madagascar*

Madagascar, a large island in the southwestern area of the Indian Ocean off the coast of Africa, is one of a group of countries with considerable biological diversity. The flora and fauna of Madagascar are highly original because of their diversity and endemism. Because of its topography, climate, soils, and hydrography, Madagascar contains a variety of formations and of ecosystems. This is shown in the map of vegetation formations (particularly forest formations) of Madagascar that was drawn up by the director of ANGAP (Association nationale pour le gestion des aires protégées) in 1995, and based on satellite data.

Madagascar has more than 4,500 km of coastline. Forest and marine diversity is high, and mangrove forests on the eastern coast constitute rich marine nurseries for shrimp, spiny lobster, holothurians, and other organisms.

The results of these new conservation measures have not always matched expectations, however. For example, one of the objectives of establishing the Mananara-North Biosphere Reserve was the integration of local populations into management plans. Emphasis was particularly placed on the participation of women. Local populations do not understand why biological diversity must be protected, however, and so this objective has not as yet been reached. Local people continue to cut down trees and to burn the forests. It is unclear whether this destruction occurs because the concept of conservation is faulty, or the conceptual and technical tools suggested were inadequate. As well, the abandonment of traditional Malagasy values may also be a contributing factor, since Malagasy culture has always been respectful of heritage, including natural heritage. Could a change of values have caused a perversion of the instruments used, and so the loss of universally recognized human and natural assets?

## ETHICS AND CONSERVATION

The role of ethics in the conservation of the resources of a country such as Madagascar is one that begs certain questions. Is the concept of ethics really applicable to conservation? Is it an overriding concept that can be liberally applied? Has conservation become an end in itself, within the context of alarming environmental degradation? Has there ever been a sense of ethics that guided actions around conservation? Who must be guided by ethics in new conservation efforts?

In Madagascar, as in other countries around the world, conservation has, perhaps, become an overworked strategy. For example, conservation has become the main objective of environmentalists inspired by the movement of deep ecology.[1] This notion – that of "pure" conservation – can be misleading, both in its global approach and in its everyday practice. Indeed, the last two objectives of the Convention on Biological Diversity – sustainable use of biodiversity and benefit sharing – suggest that the notion of conservation – the protection of biological diversity for its own sake – is out of date. In Madagascar, for example, there are three kinds of protected areas: special reserves, parks, and integral reserves. The philosophy behind the integral reserve is that of pure conservation, with no use of resources from inside its perimeters allowed. In a poor country, such as Madagascar, where citizens depend on wood for fuel, it has proved impossible to prevent use of these resources.

## CONSERVATION AND CULTURE

Saving nature for its own sake should not be the only goal of a conservation strategy. Conservation efforts also need to be associated, or be in symbiosis, with the surrounding culture – that is, the dominant cultural environment.

At present, research in Madagascar is not conducted along policy lines that help eliminate poaching or "research in a free zone." Each scientist works according to his or her own ethical guidelines, constrained by classical scientific dictates and financial requirements, and pursues research projects proposed by partners or donors. Conservation measures are not, however, simply ways in which to develop pristine, museum-like ideas of protection: instead, they must be developed along lines that benefit all those affected by the measures. Thus, researchers must be pioneers in this effort, and stringent rules of conduct must be required of them.

*An Alternative, Traditional Form of Biodiversity Conservation in Madagascar: Honey and Forest Conservation*

In Madagascar, honey is not only a culinary delicacy: it is also used as an entreaty to *Zanahary* (God). Honey used in religious offerings is unlike the honey Madagascar exports, or that the Malagasy consume daily. This "royal honey" is collected from intact forests, from natural hives in spaces within trees or between rocks. It is made from the nectar of countless forest flowers, and, as an offering, is spread over rocks in the places it is collected. This age-old belief and tradition, ensures that the forests from which honey is collected are never cut: they are considered sacred, and they provide a link to *Zanahary*. As a result, the remnant forests of Madagascar are those where royal honey is found, and the importance of this honey to the Malagasy drives efforts to conserve the relic forests that remain in the country. Economic rules of valuing don't apply: under no circumstances is royal honey sold. The spiritual value of this resource drives the conservation effort surrounding it. The importance of royal honey illustrates the strength of the link between culture and the protection of nature.

## THE BEHAVIOUR OF FOREIGN RESEARCHERS IN MADAGASCAR

Many conservation projects and programs use the services of foreign researchers. Before 1970, these researchers came to Madagascar to make an inventory of the country's flora and fauna. They came on their own or were sent by institutions such as L'institut de recherche pour le développement (IRD), or L'institut français de recherche scientifique pour le développement en coopération (ex-ORSTROM), or from other forestry and agricultural research centres.

Since 1984, foreign researchers have come to Madagascar either as permanent technical assistants or technical advisors, on short-term missions within the framework of a project, on *ad hoc* field missions, or as consultants in a specific field. Many managers of protected areas now employ foreign consultant researchers for on-site research.

In general, researchers behave well; they observe relevant national regulations, for example, and ask for authorization to gather samples or to do research. Before leaving, they table activity and scientific reports. They often give briefings and present duplicates of collected and authorized samples to the institutions designated for this purpose. During their stay, most foreign researchers instruct students, technicians, or field instructors. Many integrate easily into the population: some even learn the language and, in some cases, even a particular region's dialect.

Many foreign researchers also work co-operatively with national researchers, and their work is published within as well as outside of Madagascar.

Nevertheless, some foreign researchers do not observe regulations and do not conform to documents in force – they do not behave properly. Some take advantage of their status as researchers to export specimens from Madagascar, on the pretext of determination. These experts contribute to a plundering of the country's scientific and cultural heritage, especially when the precise destination of the specimens is not revealed. Many samples have been sold to museums abroad without the agreement or authorization of the host country, with sales solely benefiting the visiting researcher. *Patypodium* (dwarf baobab), various orchids, ostrich, shellfish, and marine and land turtles are just a few of the species that are targeted for illegal export and sale. Some researchers are dismissive of host projects, and do not maintain friendly relations with the communities they meet. Others go so far as to meddle in the political life of the country, while a few take advantage of their status as consultants to act as tourists, neglecting their research and their work. In certain cases, foreign researchers neglect to mention the names of their Malagasy colleagues in their papers, even though the research was conducted jointly. Finally, a few foreign researchers, taking advantage of the authorization they have received to do research, undertake other, unapproved research, conducted with unauthorized scientists. Such projects are often the result of a failure on the part of a researcher and a foreign research organization to submit a memorandum of understanding to the authorities in Madagascar. Scientists with institutional affiliations generally have better relationships with Madagascar authorities, local populations, and national scientists than do more independent researchers. Presumably this is because they are accountable to their institutions, and so have an added incentive to behave well and pay attention to local laws and international agreements (for example, CITES, the Con-

*Ethical Compliance in Protocols and Agreements*

International researchers working in Madagascar under IRD or CIRAD (Centre de coopération international en recherche agronomique pour le développement) are governed by protocols on scientific co-operation established between France and Madagascar. The researchers' work in Madagascar is thus described within the framework of these protocols and agreements, and they must jointly publish and distribute the results of research that is carried out co-operatively.

Researchers may make use of Malagasy data, reports, or documents for their own publications, but they must mention the partnership from which the work was generated. In general, they must also train local young people to be researchers. As well, within the limits of current texts, the sponsoring department may allow researchers to leave Malagasy territory with research material (field notes, documents, samples, or information diskettes), copies of which will have been left with the partner(s).

Both parties must consult each other every time protecting certain findings requires some discretion in publication policy. Expatriate ORSTOM personnel come under the general authority of the department of research and the manager of research programs. For duly reasoned motives, the Malagasy party may request the departure of an expatriate officer.

vention of International Trade in Endangered Species). Good practices in conservation biology cannot, therefore, rest solely on personal ethics. A professional code of ethics is indispensable, and every researcher should be compelled to conform to it. This code is a real and urgent necessity.

## THE CONDUCT OF MALAGASY RESEARCHERS

Poor behaviour, indicative of a lack of ethics or respect, is not unique to international researchers. National researchers also often display behaviour that has alienated many local residents and hindered many Malagasy conservation projects. National scientists work in rural areas, in very different cultural settings than those of the research/government institutes at which they are employed. Cultural distances exist not only between people of different nationalities, but between citizens of the same country who belong to different social groups (see Priscilla Weeks *et al.*, chapter 4). The kinds of inappropriate behaviour attributed to national researchers have included

- A tendency to adopt a moralizing attitude, indiscriminately condemning local populations because of their lack of information, education, or communication techniques. Because of such attitudes, some rural target communities now believe, for example, that the lives of lemurs are more important than the lives of their children. Villagers thus become hostile and even initiate acts of sabotage. During the five years of the Environmental Action Plan, many such errors have occurred. In 1984, Madagascar had a National Strategy on Nature Conservation. In 1987, this strategy became an Environmental Action Plan, the first phase of which has been completed. The second phase has begun. Conservation measures on site preservation have been developed through the establishment of approximately 50 protected areas.
- Haste and hurry in research and study operations in the field. Donors often demand reports within very short time frames, and scientists accept these deadlines. This is not conducive to quality research. Any field study undertaken with a concern for ethics should include enough time to gain the confidence of the partners involved.
- The creation of an impression of stolen knowledge, and of a violation of social intimacy. Such impressions are created when information is taken from local peoples, and when the donors do not receive any form of return on their donation. This impression can be very strong and causes a permanent strain on general researcher/community relations, which can last for a very long time. Rogelio Cansarí (chapter 3) discusses this point in depth.
- The absence of any practical, multidisciplinary approach. The necessity for such an approach is preached constantly, yet in most cases, each researcher works within his or her own area of expertise and does not communicate with others.
- No communication of research conclusions to the people who provided data and information. Often, there is little effort made to distribute and simplify, and hence ease the retention, of information (see Rogelio Cansarí, chapter 3, for a fuller explanation of this point).

*Rights and Obligations of National Researchers*

National researchers are governed by a new act (Decree # 96/728, 21 August 1996), and are bound by a certain number of obligations and duties to compensate for the rights and new privileges they have received. They are thus called upon to observe research ethics and rules of conduct. They also have service obligations, particularly full employment, punctuality, attendance, and honesty.

Besides the rules regarding professional secrecy contained in the Penal Code, researchers are also bound by professional discretion concerning any documents, facts, and information brought to their attention as part of the performance of their duties, and are not allowed to use these documents outside of the activities related to their professional duties.

Any invention or discovery made by a researcher working in a national research centre as part of the performance of his or her duties belongs by right to the research centre, which alone may apply for any related patent(s), in Madagascar or any other country. The patent may bear the name of the inventor. Should there be any commercial production of the patented item, the establishment may approach the inventor. If the national research centre declares that it has no interest in the invention, the inventor is free to proceed. When research is conducted by employees of

- The willingness in times of increasing poverty, to sell specimens and send them as scientific samples to dealers.

In developing countries, there is an undeniable link between environmental conservation and development actions. Behaviour that is positive and laudable has been noted over the last decade, following the calling into question of the concept of development. Development is now perceived in terms of its sustainability, even though at first glance the two concepts appear to be in opposition. Development programs and projects have been carried out using participative management and are based on the social and human sciences (see Wigley and Baser, chapter 7). Equal importance is given to human beings and to the natural resources that they must manage and use.

the national research centre together with researchers of other organizations under contract with the centre, the contract will determine the ownership of inventions. To compensate for these obligations, the State gives these employees the right to unionize and the right to strike. They may also become involved in money-making activities in line with maintaining the honour of the profession (expertise and consulting) after having obtained the permission and protection of their line supervisors. They may not be harassed as a result of scientific and educational activities undertaken for the centre, and are protected against any threats or attacks during or because of the performance of their duties. The State must make amends for any hardship endured, unless the interested party is at fault in a way not related to the service.

Ethics do not include moral sanctions. Nevertheless, a certain number of disciplinary measures have been developed for research personnel as a realization of ethical transgression. These include warning, blame, removal from the promotion chart, loss of seniority, demotion, mandatory retirement, and dismissal with or without pension rights.

Proper ethical conduct and exceptional service to the nation as part of the performance of their duties will earn researchers the following: congratulations, seniority gains, immediate promotion, emeritus, and honorary status.

## WORKING TOWARDS AN ETHICAL FRAMEWORK

The primary element of conservation ethics is the development of respect for cultures, traditions, local peoples, and interlocutors – in short, all potential participants in the conservation effort. Notions of conservation, as well as conservation actions, will therefore be different in each country or region (see Priscilla Weeks *et al.*, chapter 4, for further discussion of this point).

The following attitudes and actions will help develop positive relationships between all participants in conservation efforts:

1 The recognition of traditional authorities as allies and facilitators in every conservation measure.

*The Role of Women in a Science Aimed at Serving Development*

The political context in Madagascar in the 1990s has not allowed university workers to develop and deal with their professional obligations in peace. Science has been given a minor role and has not, for the most part, been a part of development efforts.

On the other hand, widespread poverty, the increasing number of female scientists, and the growth of a pro-female philosophy have contributed to the birth of Women and Science in Madagascar, which brings together female scientists from both Malagasy universities and national research centres. This nonprofit association aims to give new life to science in Madagascar by breaking down institutional barriers and promoting group work by making research results available to the general public. It aims to do this in as simple a manner as possible in order to internalize results among people and developers as a group while promoting individual researchers. Renowned female scientists are very rare in Africa. The aim is thus to increase their number in the long term by encouraging female specialists to undertake scientific and technical studies and by organizing promotional seminars, educating more women, and making them responsible for their own well-being.

2 The inclusion of those who understand Native traditions and lore.
3 The recognition of existing Native biological knowledge in official and institutional statements. In Madagascar, this action is already being taken – traditional practitioners are now becoming involved with research activities in the pharmaceutical field, something that would have been inconceivable just ten years ago.
4 The enumeration and follow-up of local customs in the field. This will ensure not only that operations run smoothly, but that social order is maintained and that neighbourly relations are developed.
5 The use of local dialects. The role of language in conservation policy is important – communication is key to conservation efforts.
6 A universal manual on proper conduct, which emphasizes respect for and recourse to local cultures, should be optimized and adopted in all areas of research and by all researchers working in the field.

At the University of Antananarivo's faculty of science, many female teacher-researchers do not have children, even if they are married. These women have thus given their professional lives priority over their personal lives. This trend seems to be diminishing if not disappearing, however, in the younger generation of female scientists (those born after 1969). Today, for example, many women conduct their laboratory or field research throughout their pregnancies. One female researcher, who was four months pregnant, conducted a 45-day field trip to study the production of rice and the exploitation of forest resources by villagers. During this time, she visited – on foot – villages scattered around a 10,000-hectare reserve. She finished her official report before she gave birth, and is now considered a symbol of courage and rigour among young female Malagasy scientists.

Female scientists in Madagascar do not confine themselves to academic research positions: they also occupy strategic and decision-making positions at the national level. Women are responsible for the politics surrounding biodiversity in Madagascar, and serve as directors in various government ministries ...

Despite these advances by Malagasy female scientists, however, few have succeeded in breaking into the international scene as representatives for Madagascar.

Although it remains to be proven, female researchers seem to be more adept in developing positive relationships with local people, because they are generally more attentive, careful, cordial, and affable. Nevertheless, Malagasy women are at a disadvantage in some regions because it is considered inappropriate for them to speak in public, which hinders their work considerably. In such instances, ethics would require them either to conform to tradition or to bring along male colleagues.

CONCLUSION

As we have seen, despite certain particularities the difficulties that conservation projects encounter in Madagascar reveal the same tensions that exist in many tropical countries: between hard-line conservation and more flexible protection, between the traditional notion of development and sustainable development, between national policies and the actions of researchers, and between researchers and local populations. Madagascar might therefore serve as an example, to be used as a basis for a general code of ethics for researchers working in developing countries.

This chapter emphasizes that ethical problems occur for both national and international researchers. Neither Malagasy nor foreign scientists can consider themselves masters of nature or owners of natural resources. All scientists must recognize the potential contribution of local peoples to conservation efforts, and consider conservation as an ongoing process, one that is integrated into the lives of local populations.

Regardless of their national origin, scientists must remain humble, honest, and understanding when others are involved in conservation efforts. In short, they must respect everyone who may be affected by their actions and decisions. This is the price that must be paid in order to ensure that conservation projects do not simply save "dying" things that can be managed, and that conservation ethics will become a fundamental element of all projects.

NOTES

1   A. Noess, "The Shallow and the Deep, Long-range Ecology Movement. A Summary," *Inquiry* 16 (1976): 95.

## 6 Conservation Biology and Environmental Values
### *Can there Be a Universal Earth Ethic?*

BRYAN G. NORTON

EDITORS' INTRODUCTION

The opposition between preservationist and conservationist attitudes toward biodiversity has been extensively discussed. In North America, this debate underlies the tension between "deep ecologists" and "utilitarians." In chapter 4, Lala H. Rakotovao and her colleagues show that this tension has spilled over to developing countries, where it adds to the difficulties of protecting biological diversity. In this chapter, Bryan Norton champions and advocates an alternative to the quest for a middle ground between "intrinsic" and "utilitarian" values of biological diversity. He explores new dimensions of the problem within the specific context of the Earth Charter, and argues that "placeless evaluation" should be rejected, since biodiversity has different value for different people at different times and in different places. He calls for the acceptance of an array of cultural perceptions, echoing the suggestion made by Lala H. Rakotovao that the loss of traditional Malagasy values might partly explain the destructive attitudes of some of the present Malagasy generation.

INTRODUCTION

Nature and species living in the wild are valued differently by different people and cultures, and these differences can produce con-

flicts between conservation biologists and indigenous populations
that inhabit biologists' study areas. These conflicts can and have
led to misunderstanding and disputes, and it is important to
understand to what extent conservation biologists, acting on their
own values, should press their judgments, defer to differing values
of inhabitants of countries in which they study, or seek compro-
mise. How this problem is dealt with may in many situations
determine the success or failure of research projects and protection efforts.

The task of articulating a professional ethic for conservation
biologists – the task of this book – differs substantially, of course,
from the task of constructing a viable theory of environmental
values. I believe, however, that the former task would be more
manageable if some level of consensus were reached on the
nature of environmental values and valuation and on the key
terms and measures relevant to the study and protection of biological diversity. The purpose of this chapter is to explore these
latter, background issues, and to do so by discussing the concur-
rent efforts to create an Earth Charter, something which is
intended by its proponents to articulate a shared ethic for the
protection of nature. An examination of this attempt may clarify
the context in which we search for a professional ethic for conservation biologists.

Recent international discussions of biodiversity policy have
established two points: (1) that there is a growing international
commitment to sustain and protect biodiversity, and (2) that there
is little agreement regarding *why* this should be done. Thus, while
a significant international consensus on policy has apparently
emerged, this consensus is not yet grounded in a consensually
accepted value theory that explains why biodiversity protection,
however strongly supported, should be a high priority for environmental policy. Lack of agreement on the second point led to dis-
agreements at the Earth Summit in Rio de Janeiro, for example,
where delegates disagreed about whether to emphasize nature's economic, "utilitarian" value or its "intrinsic" value, which is defined
as value that exists independently of human values and motives.

The debate between saving nature for future use and saving
nature for its own sake has infected the discussion of a proposed
Earth Charter, which is being urged as a next step in the development of a legal and political framework to guide local, regional,

national, and international efforts to protect nature.¹ In an early
statement of principles, Steven C. Rockefeller, an advocate of the
Charter, stated that

In order to address the many interrelated social, economic, and ecological
problems that face the world today, humanity must undergo a radical change
in its attitudes, values, and behavior ... The purpose of the Earth Charter
Project is to create a "soft law" document that sets forth the fundamental principles of this emerging new ethics, principles that include respect for human
rights, peace, economic equity, environmental protection, and sustainable
living.²

He goes on to say that he hopes the Earth Charter "will become
a universal code of conduct" that expresses "the shared values of
people of all races, cultures, and religions."³ These goals are laudable, and an Earth Charter is a wonderful idea. But is it possible to
share Rockefeller's optimism about the early arrival and consensual
acceptance of "this emerging new ethics"? *Is* there an overarching
ethic that adequately represents the values of all peoples? How
might one articulate such an ethic, given the existing tensions surrounding evaluations of nature? This chapter explores these questions by surveying the oft-cited value theories – utilitarian/economic
and nonanthropocentric – and by arguing that neither of these
approaches is likely to provide a unifying ethic that could support
a universal Earth Charter. The chapter points toward a new
approach to environmental evaluation, one that is more likely to
lead to an inclusivist ethic.

One reason this debate has been so frustrating and polarizing is
that we have been asking the wrong questions in the wrong way,
given the task at hand. As well, neither of these theories provides an
inclusive ethic that can guide actions affecting nature, including
wild life forms. Part 1 examines an alternative to Economism and
Deep Ecology, an alternative that can best be explained by showing
how these two opposing positions on values actually share important and highly controversial assumptions. In Part 2, the problem
of choosing the most important targets of environmental protec-
tionist policies and priorities is discussed, and the role of values in
conservation policies is examined. In Parts 3 and 4, a sketch of one
alternative approach is constructed, an approach that may well
provide an inclusive theory of environmental values.

PART I: VALUE THEORIES AND
BIOLOGICAL DIVERSITY

So far, most discussions of the process of evaluating nature have been based on two theories: Economism and Deep Ecology.[4] Although both theories have many variants, this chapter will examine their most general forms.

Economists believe that environmental values are economic values. According to this theory, elements of nature have instrumental value only, and should be valued like other commodities. Of course, Economists recognize that there are no existing markets for many environmental goods and services, so methods other than market-behaviour measurement must be used if we are to correctly describe human preferences vis-à-vis the environment.

Deep Ecologists and other non-anthropocentrists directly oppose this instrumentalist, preference-based theory. They argue instead that some elements of nature have inherent value, and that these elements are therefore deserving of preservation for their own sake. According to this view, human individuals and some other elements of nature – individuals, species, or ecosystems – have their own, intrinsic values, values that are not dependent upon the preferences of human beings.

Interestingly, the tension between Economism and Deep Ecology is openly expressed within Rockefeller's "Summary of Principles." One principle of the new Earth Charter (found in the "Worldview" section) states that "Every Life form is unique and possesses intrinsic value independent of its worth to humanity. Nature as a whole and the community of life warrant respect." Meanwhile, the Charter's first principle on sustainable development states that "The purpose of development is to meet the basic needs of humanity, improve the quality of life for all, and ensure a secure future." Are these principles consistent? It depends, of course, on how the terms are defined. If, for example, we assume that "development" means development of natural resources, including biological resources, and that, as the syntax of the principle implies, "all" refers to all humans, then the statements are indeed inconsistent. Whether or not these principles are incompatible, they certainly express a tension between two broad ways of valuing nature. The first principle emphasizes the valuation of nature apart from humans and their activities, whereas the second principle emphasizes the evolution of nature insofar as it fulfills human needs. Subsequent drafts of the Charter have continued to define both the explicitly "intrinsic value" inherent in nature, and the human rights aspect of sustainable development.[5]

The tension between these two value theories also tends to polarize discussion of international efforts to protect biodiversity. Environmentalists from the United States and other developed countries espouse intrinsic values in nature, even though the U.S. government, because of economic concerns, has failed to ratify the Convention on Biodiversity. It is common for spokespersons from the developing world to complain in international policy forums that First-World countries have already exploited and converted their forests; now they ask *us* to forgo forest-based development and attendant increases to human welfare. Even as governments of developing countries are attempting to maximize economic development based on exploitation of natural resources, there have emerged minority groups, including minorities from Indigenous cultures, that are opposed to economic exploitation and have attempted to retain or resurrect their animist religions as a counterbalance against this exploitation. The tension over why and how to value nature therefore has real consequences, and makes it more difficult to forge coalitions between Northern and Southern countries.

More importantly, the lack of consensus on foundational values also affects hopes for the development of an Earth Charter: it becomes virtually impossible to embody the use-and-development ethic, its attendant utilitarian value concepts, and the save-nature-for-its-own-sake ethic in one document. Given the tension between the two sides in the debate, development of a more inclusivist ethic will require both sides to move toward a middle ground, something which has not yet occurred.

It is important to recognize that these opposing theories rest on a cluster of highly vulnerable assumptions. Both Economists and Deep Ecologists accept a sharp dichotomy between values that are inherent and those that are instrumental; furthermore, both groups use this dichotomy to separate nature into beings or objects that either have or do not have "moral considerability." In a particularly strong variant of economism, for example, Gifford Pinchot (first head of the U.S. Forest Service), said that "There are just two things on this material earth – people and natural resources."[6] Pinchot was thus enforcing a sharp dichotomy between human beings and the natural world. Only the well-being of the former, he believed, should be taken into account in our determination of acceptable behaviour. Interestingly, the position of Pinchot and his economist allies is consistent with that of Immanuel Kant, who, it is commonly believed, was opposed to the consequentialist emphasis of utilitarians. For Kant, only rational beings could be "ends-in-themselves." Both Pinchot's utilitarianism and Kant's rights theory are thus based on a sharp distinction among entities – those regarded as being "ends in themselves" and those objects that can be used, without restriction, in service of those ends.

Economists and Deep Eecologists, then, agree that there must be some special status for those beings that have non-instrumental value: they simply disagree over which objects in nature actually have this special status. For economists such as Pinchot, special status is co-extensive with humanhood; for the Deep Ecologists, moral considerability is co-extensive with a much larger subset of nature's components. Either way, the sharp distinction between instrumental and inherent values ensures that questions of environmental value are posed in all-or-nothing terms. For the Economist, Should we protect this river? becomes Does this river have net positive economic value (for humans) or not? For the Deep Ecologist, Should we protect this river? becomes, Does this river have inherent value? These questions are usually formulated in incommensurable theoretical frameworks: what they share is their tendency to elicit yes-or-no determinations of the value of objects, because valuation is considered a function of categorization. The value of a thing is a consequence of the kind of thing it is – the moral/evaluative category within which it is correctly placed.

Another, related consequence of the bipolar formulation of environmental valuations is the apparent bias of both sides in favour of evaluating objects or entities, rather than evaluating dynamic processes and changes in processes. Protection is assumed to mean protection of items in an inventory: should we try hardest to save genes? Individuals? Populations? Species? Ecosystems? This object bias is, of course, endemic to all of Western culture, at least since the classical period. It represents the triumph of Plato's concern for constancy of forms that constitute reality over the ideas of Heraclitus, who, around 500 B.C., declared that "All is in flux."[7] This ideological triumph also paved the way for modern scientific reductionism, which seeks explanation in the motion of elementary particles.

The "atomistic" approach, which emphasizes elements, is so deeply engrained in Western thinking that alternative conceptualizations of nature have only been considered relatively recently. Since the publication of Charles Darwin's *On The Origin of Species*, however, the importance of systemic change and irreversible developments – of complex, dynamic processes – has asserted itself.[8] This revolution has also extended to physics, and physicists are now leaders in an interdisciplinary effort to develop a more dynamic world view, evidenced by the ever-increasing emphasis placed on non-equilibrium dynamics. The full implications of this world view are just now being felt, and it may be decades before these concepts are well understood, but creative work in non-equilibrium system dynamics is already leading to new insights in ecology, and this direction holds promise for environmental policy.[9] We do know that the full absorption of evolving systems thinking into environmental management will have a far-reaching impact on advocated policies, and will almost certainly require more attention to interspecific relationships and system-level characteristics.

Another similarity between Deep Ecologists and Economists is that they are both looking for a universal, "monistic" approach to values. Monism, as defined by Christopher Stone, conceives the ethical enterprise "as aiming to produce, and to defend against all rivals, a single coherent and complete set of principles capable of governing all moral quandaries."[10] Commitment to monism explains the categorical, all-or-nothing character of the two competing theories. Both Economists and Deep Ecologists believe that there is just one kind of ultimate value; they differ only in how widely they find that value. It seems unlikely that either of these monolithic theories of value will prove rich enough to guide difficult, real-world choices about what species should be saved, where conservationists should concentrate efforts, and how they should set priorities. But should that not be precisely a theory of environmental value's role in the conservation policy process?

PART 2: RETHINKING THE PROBLEM OF
CONSERVATION PRIORITIES AND TARGETS

Interestingly, the same entity-oriented concepts that have divided environmentalists over value theory have also affected practical policy thinking. For example, suppose one asks, What should be the highest priority in protecting biological resources? Typical answers to this question, given the Western tendency toward an entity orientation, might be, "all species" or "all species and all ecosystems." Other answers might identify particular types of species – for example, "producers," as the most important. Note how easily we fall into providing a list of entities – an inventory of things that should be saved. The establishment of conservation priorities is thus accomplished through ranking various categories according to their perceived importance.[11]

Like Economists and Deep Ecologists, biologists tend to talk about conserving inventories of objects, and setting priorities among these. Whereas discussions of value theory emphasize philosophical considerations of what has ultimate value, however, in a management context these pure, philosophical considerations are inevitably mixed with questions of methods and means, and even with recognitions of political constraint. For example, a conservation biologist might believe that species are the highest priority, and so advocate policies that preserve ecosystems and habitats because the systems approach is the most efficient means to achieve the goal of saving species. Conversely, a policy analyst might argue that we should save ecosystems for the future, but that the best way to do this is by legislating the protection of species (because they are, for example, fairly easily counted). The policy analyst and the conservation biologist working in the field differ, therefore, from the moral theorist because they must take into account empirical realities imposed by the intricacies of systems, by lack of knowledge, and by political constraints.

Although the differences between these priorities are very important, both the philosophical values debate and the practical policy debate are framed as questions that can be answered by presenting lists of entities. Setting priorities, given this entity-and-category orientation, therefore, is a matter of *excluding* some kinds of objects from a "preferred" list.

To understand how this entity orientation has affected policy discussions, it is useful to briefly explore the history of policy debate in the United States over biological resources. Attempts to protect biological resources in that country can be divided (somewhat arbitrarily, but usefully) into three phases. As early as the seventeenth century, local shortages of important game species such as deer led to "bag limits" and other restrictions on the taking of those species.[12] As human populations increased, more and more species suffered population decline and more and more restrictions were contemplated and then imposed. These restrictions were sometimes successful and sometimes were not.

What is important in this process, however, was the formulation of the policy issue.[13] In this first phase of biological resources protection – the "single-species" phase – the protected object was always a species or the populations of a species. This first approach continued well into the twentieth century, as the populations of more and more species underwent decline in response to human exploitation and habitat conversion. The approach culminated in the Endangered Species Act of 1973. This act, which was unquestionably a remarkable advance in policies protecting nature, nevertheless exemplified the original understanding of the problem as one that should and could be addressed at the level of species.[14] Accordingly, success in protection efforts meant the protection of an inventory of existing objects.

By the early 1980s, a number of scientists and policy analysts began to explicitly question whether the species-by-species approach was sufficiently comprehensive to protect all biological resources. The term "biological diversity" came into common use in the early 1980s, and it was subsequently shortened to "biodiversity" in conjunction with an influential Symposium on Biodiversity, which was sponsored by the Smithsonian Institution and the National Academy of Sciences.[15] This symposium ushered in the "biodiversity phase" of policy debate on biological resources. The new term, as used at the symposium was defined as "the sum total of distinct species, genetic variation within species, and the variety of habitats and ecological communities."[16] While this new concept no doubt represents a significant advance in the search for more comprehensive policies to protect biological resources, it is nevertheless problematic. The definition still treats protection of biological resources as the protection of objects – an inventory of items. In this sense, biodiversity is simply an extension of the single-species phase: it expands the list of items, but does not depart from the goal of maintaining an inventory.

Despite this formal consonance with older approaches, the definition of biodiversity introduces a more dynamic conception of resources. The third element of the definition – "the variety of habitats and ecological communities" in which species exist and adapt

– refers to a whole range of ecological processes, although it is listed as elements in an inventory. This change, which recognizes that species actually trace multiple and changing trajectories through time, has the subversive effect of undermining the comfortable assumption that protection of biological resources means protecting a static list of entities. While the term biodiversity is useful in expanding the focus of protection beyond species, and is useful in some contexts, it is ill-suited by its entity orientation to capturing the essential features of the emerging process orientation of biology and ecology.

This inadequacy is now leading to the development of a third phase of goal formulation, although it may be too soon to discern the details clearly and considerable controversy still exists over its development. One thing seems certain, however: in the new phase, processes will be more important than entities. We might therefore call this emerging phase the "ecosystem processes" phase: since it is unclear how processes will be described in the future, it is also unclear how environmental evaluation will be expressed in a more process-oriented mode. Some biologists have suggested that medical analogies are helpful in redirecting our thinking toward process evaluation, and use terms such as "ecological health" or "ecological integrity."[17] Others have argued that these analogically derived terms have no clear scientific meaning, and illegitimately mix description and evaluation.[18] We can avoid the purely linguistic issues by arguing for more linguistically neutral conclusions: in the future, our descriptions of nature will become more dynamic and process-oriented, and our evaluation of biological aspects of these processes will have to change to accommodate these changes in description.[19]

The trend away from static and toward more dynamic models of environmental problems and goals has also put into focus a corresponding trend toward more wholistic and systems-oriented managerial models. In the United States, increasing emphasis has been placed on "ecosystem management plans," and often involve "adaptive management." These planning and management processes – which often involve all levels of government, multiple agencies of government, and the public – are applied to systems bounded by natural features of the landscape.[20]

In the United Kingdom, a similar trend toward holism is made clear by planners and political theorists such as Michael Jacobs, who argue that making economics "ecological" is not enough, and that it is necessary to adopt an even broader analysis of environmental problems – "socio-ecological economics."[21] Environmental policy experts and advocates on the European continent have, using several technical vernaculars, similarly advocated more holistic and integrated environmental management. One group, building on the ideas and practices of Impact Assessment, has advocated what is now being called "strategic environmental (impact) assessment."[22] Proponents of this approach extend traditional impact-analysis techniques to larger systems and advocate applying their tools not just at the project level, but also to policy impacts and even proposed legislation. This approach, still in its infancy, advocates the use of impact assessments at project, local, regional, and perhaps even national and international levels. The embedding of many smaller, simultaneous impact studies in a multi-layered, integrated assessment represents one step towards a more holistic environmental management.

Another promising step has been championed by Dutch theorists, modellers, and practitioners, some of whom are developing the model, TARGET, which includes both social and ecophysical features. The idea behind this approach is that debates and controversies surrounding new technologies provide informal methods of technology assessment, and that such informal processes may be integrated into larger, multi-scalar models of humans and nature.[23]

These trends all point toward a more holistic approach to environmental management and (at least implicitly) a greater respect for dynamic processes, all positive steps. The remainder of this chapter examines in greater detail the construction of a process-oriented theory of values and a practical, process-oriented method of evaluation as elements of a larger theory of adaptive ecosystem management.

PART 3: ADAPTIVE MANAGEMENT –
AN ALTERNATIVE TO MONISM

Considerable progress has been made in understanding environmental problems as problems of adaptation within complex, multi-scalar, dynamic systems. This trend may emerge as the management approach of the twenty-first century, encouraging a more holistic and process-oriented view of conservation. C.S. Holling and his colleagues, who have been described as adaptive managers, have argued that large, landscape-scaled ecological systems tend to become "brittle" under continuous exploitation and that these large systems can disintegrate and then gradually "re-equilibrate" at different levels of functioning or with quite different structural organizations.[24] On the basis of this hypothesis, and with some success in modelling resource management in the field (including work on spruce budworm outbreaks and fisheries management), Holling and his colleagues have become the champions of dynamic, non-equilibrium modelling in resource management. They argue that environmental management cannot be modelled in single-equilibrium systems, and that the impact on natural systems can approach thresholds which, if exceeded, can cause discontinuous and rapid change into an alternative steady state.[25]

It has been often noted that such large-scale changes result in systems that use human services less productively, or are less attractive to human users. This judgment, of course, cannot be made in purely biological terms because it presupposes that social values are its basis. Adaptive management, according to this argument, would choose exploitational patterns that mimic natural processes in order to minimize the likelihood of accelerating system-level change and the subsequent loss of human use. On a theoretical level, this insight is embodied in the argument that states that good management must care for both the productivity and the resilience of the ecophysical system. "Resilience" is introduced as a measure of the magnitude of disturbances that can be absorbed before a system centered on one locally stable equilibrium "flips" to another equilibrium.[26]

This hypothesis has also been elaborated through the explicit use of concepts from "hierarchy theory," an application of general systems theory, to ecological modelling. Hierarchy theory models ecological systems according to two assumptions: (1) that all observations and measurements must be taken from some perspective within a hierarchically organized dynamic system; and (2) that the systems, as modelled, exhibit nestedness, with smaller subsystems changing more rapidly than do the larger systems that form the subsystems' environment.

Given this framework, it is possible to model environmental problems in a multi-scalar system.[27] At the individual level, organisms (including humans) survive by responding creatively to a set of opportunities and constraints that are presented to them by the environment they inhabit. The range of available options is a function of the structure and consequent functioning of the habitat. Individuals' successful choices are encoded into the system as information about what "works," given current system organization. Thus, adaptation means to act successfully on information flowing from the environment to individual actors. If the first generation clear-cuts all of the available forest, for example, opportunities are lost and attention shifts to constraints experienced by the second generation, such as how to survive despite an inadequate wood supply. Information also flows upward in the system, because the collective choices of many individuals in a nested system gradually alter the context in which the next generation faces adaptation. The nested subparts also constitute the larger systems, and exhibit a cumulative, slower-scaled impact on the structure and function of the systems that support human choice.

Writers and practitioners of this tradition have successfully developed an adequate characterization of how environmental problems emerge as problems of adaptation at the individual and cultural levels. This multi-scalar system allows the conceptualization of both an "individual" and a community scale of action upon which populations interact with their environments. One important consequence of this multi-scalar and multivariate formulation is that neither means nor goals of sustainability can be concretely set in the beginning, and the quest for sustainable human communities must involve many individual processes of experimentation, revision of scientific understanding, and reformulation of community goals. This adaptive, experimental approach requires the careful design and nurturing of institutions that are capable of fostering social learning and equitable, participatory decision making.

In the past few years, adaptive management work has virtually merged with the work of "ecological economists," who have broken with mainstream, welfare economics on several important grounds.[28] Ecological economists have rejected the idea of "weak sustainability," popular among mainstream economists, that a generation acts responsibly toward the future if it adds to, rather than diminishes, the total stock of human capital. If we accumulate wealth – that is, assets in any form – then the future will have an adequate investment and opportunity base that is equal to that of the present. In other words, the next generation will be just as well off as we are, according to mainstream economists and weak-sustainability theorists. Ecological economists go beyond this generalized obligation, however, arguing that there are specific features of the environment – "natural capital" – that should be preserved in trust for the future.[29]

Adaptive managers and ecological economists have thus joined forces, arguing that the resilience of ecological systems is surely one important example of natural capital. These combined forces argue that system resilience is a more useful index of environmental sustainability than alternative measures, such as economic growth measures or simple carrying capacity measures, because economic activities are sustainable only if the life-support ecosystems on which they depend are resilient.[30]

Adaptive management and ecological economics have therefore provided a theoretical model for understanding environmental problems as (what could be called) "cross-scale spill-overs." For example, if an individual cuts down one tree, or even an entire woodlot, this action will have little long-term impact, provided other individuals let their trees grow. If all forest-owners in a watershed, however, clear-cut their forest within a few years, a standing resource – an option for use – will have been eliminated for decades. There may also be indirect effects, such as soil erosion and reduced stream-water quality, that can last much longer.

Environmental problems can thus be understood as multi-scalar, or cross-scale spill-over problems – an idea that is explored further in the next section. Adaptive managers have provided a working, holistic model that illuminates the emergence of environmental problems at the systems level and that integrates human, cultural activities into the larger-scaled, and, normally, slower-changing landscape. Environmental problems emerge at the system level because of the impact of increased human populations, technological power, and consumption. Cumulative individual choices can accelerate change, cross crucial thresholds, and cause systems to undergo rapid structural reorganization. If Holling and his colleagues are correct in asserting that flips into new states will result in habitats that are less productive of humanly valued goods, then these changes will be experienced by individuals in subsequent generations as a constriction of the survival options available to them. The mix of opportunities and constraints presented by the habitat will have shifted for the worse.

It is now possible, following Holling and the adaptive managers, to think of values as emerging within a dialectic between culture and nature. Each generation faces a mix of opportunities and constraints, and the cumulative choices of each generation will affect the mixture of opportunities and constraints that coming generations will face. This same set of concepts can be used to characterize and analyze intragenerational justice in international situations. For example, if a particular development path is imposed upon developing nations by developed nations – through colonization or, more recently, through economic hegemony – and that development path reduces options for resource use for Native inhabitants of a region or country, then a problem of intragenerational justice arises.

Increasingly, development experts agree that long-term sustainability of resources depends, within nations, on the rapid development of fair and equitable institutions that provide access to those resources by all segments of society. Resource protection will also require greater international equity and a truly sustainable world can only emerge if the Northern nations reduce consumption and encourage greater equality of access to resources in Southern nations. An obligation to live sustainably implies an obligation to build, and to help less developed nations to create, open and equitable institutions for controlling and ensuring fair access to resources and to the opportunities they represent. In this sense, the goal of creating institutions and policies that will guarantee fair access to resources for future generations presupposes the immediate goal of building better, more equitable institutions (on local, regional, national, and international levels) that provide access to resources.

Intragenerational justice therefore depends upon the achievment of greater intergenerational justice in the near future. Despite the importance of intergenerational justice, however, it will not be discussed in detail because the focus of this chapter is the sustainability of biological resources through the study and actions of conservation biologists. This focus implies a multi-generational, temporal frame for our analysis and directs attention toward intergenerational expressions of the ideal of fairness.

Long-term survival as a community or culture requires, in the short term, adequate economic opportunities and a reasonable pace of economic development. In the long term, it means that future generations must have a roughly equal or superior mix of opportunities and constraints. Thus, if measures of economic activity such as the gross domestic product (GDP) are to be used as comparisons of welfare over time in the short run, then what is also needed is an "opportunities-constraints Index" (OCI) to measure changes in opportunities available in the future. The OCI could compare rising or falling options and opportunities that the environment presents to actors across intergenerational time frames. It can therefore rate development paths (packages of policies and choices) according to their impact on the range of free choice (opportunities to adapt) that will be open to posterity.

Thus, a good policy or program would, when added to the current economic and ecophysical dynamic, be expected to (a) increase economic welfare (on the scale of months and years), and (b) maintain a non-declining stock of resource-based options for future generations – that is, it must perform well on more than one temporal scale. Adaptive management therefore provides a useful, and quite general, representation of many environmental problems as cross-scale spill-overs or impacts of cumulative human activities. Because this representation is expressed in a multi-scalar, dynamic system, it also sets the stage for a more dynamic approach to the evaluation of change and processes of change.

Adaptive management has been given a political component by Kai Lee, who accepts the characterization of environmental management as an experimental, community-based search for ways to exploit natural systems without undermining their healthy functioning.[31] Adaptive management, he says, should represent a negotiation within a politically organized ecosystem management process. Scientists, working within a political process in which stakeholder groups express and defend their interests, attempt to develop trust sufficient to undertake management "experiments". These experiments are designed to produce both an epistemological "community" devoted to experimentation and social learning, and a reduction in uncertainty and related situations. Through this process, the community guides scientists to study aspects of the system that are important to the community, and scientists respond with studies that will help reduce the negative impact of valued human activities.

Although this general model is laudable, I believe current multi-generational models are put forward as descriptive models only.

Values, if they are mentioned at all, are treated exogenously.[32] Current versions of adaptive management theory, economism and deep ecology, assume that environmental values have a source and are fully determined outside of the policy process. Although they are treated exogenously, valuations by individuals are important system drivers because individual behaviour expresses preferences that result in the cumulative impacts that threaten resilience. As long as values are maintained as exogenous and hence independent variables, the system of management can offer no remedy or informational feedback loop if individuals in the community express preferences that promote more and more negative cross-scale impacts.

This problem can be seen if one simply observes the virtual merger of adaptive management with ecological economics. Ecological economists set out to define "natural capital" as those features of an environment that are essential to protect future welfare. Ecologists suggest that changes in ecological organization – flips into new states – are socially disvalued. Resilience is defined as a descriptive characteristic of ecological systems, however, prompting us to ask if we should protect the resilience of systems. We cannot answer, as the welfare economists do, that individuals, taken in aggregate, "prefer" to protect systems, if the current pattern of preferences drives economic development and threatens resilience.

What adaptive managers and ecological economists want to say is that humans should prefer a higher or over-riding value – our obligation to protect resilient ecosystems. This alternative cannot be adequately expressed, however, even within an enriched vocabulary that includes the concepts of ecology (which describes changes and thresholds in the ecological system) and of economics (which describes preferences as they are expressed by present consumers). What is lacking is a theory of value that (a) establishes the possibility of articulating values that can compete with currently felt preferences, such as an obligation to sustain opportunities for the future, and (b) some way to link those long-range values to physical features such as resilience. Lacking such a value theory, adaptive management can simply describe a system that is going haywire: it can offer no analysis of how the system might right itself by affecting the driving, independent variable – current preferences. Preferences, and the evolution of preferences, must be a part of the adaptational process; adaptive management must test and revise

our values as well as our empirical hypotheses. Individual preferences and social values, as well as the institutions that shape them,
must be considered and modelled as endogenous to the social
process of environmental management.

PART 4: ALTERNATIVES TO MONISTIC
ASSUMPTIONS AND THE ENTITY ORIENTATION

This chapter has argued that the philosophical debate over inherent
value in nature and the policy debate over how to set conservation
priorities have both presupposed an entity orientation. This is in
keeping with an unquestioned assumption of virtually all of
Western thought. It has also been argued, however, that an entity-
based evaluative scheme, even if it were available in operational
form, would soon be rendered obsolete by recent developments in
environmental management. Management thinking is moving away
from single-species management and will likely eventually move
away from the inventory-of-objects approach altogether, because
environmental problems, as has been shown, are best understood as
problems of adaptation across multiple scales of time.

Part 4 examines the prospects for a process-oriented system of
environmental evaluation and explores some of the general features
of such a system by discussing the consequences of denying the
crucial, shared assumptions of both Economists and Deep Ecologists. Once freed from the shared assumptions that bind both viewpoints in polar opposition over classifying objects of value, it is possible to look with fresh eyes at questions of value and policy. The
effects of rejecting the central assumptions shared by these opposed
groups will therefore be discussed.

*1 Denying the dichotomy between instrumental
and intrinsic valuing*

It has been shown that Economists and Deep Ecologists share a
complex web of beliefs about the nature of environmental value.
The defining feature of that web is the belief that a sharp distinction must be drawn between two kinds of value – intrinsic and
instrumental – and two types of objects – those that are valued for
their own sake and those that are not. J. Baird Callicott argues that
the word "value" is a verb, not a noun, and that valuings are

always acts or dispositions of conscious beings.[33] Using this interpretation, it is possible to consider a range, or continuum, of ways of valuing. Each way represents the decision of a conscious valuer who experiences nature in specific situations. Given this adverbial conception of environmental valuation, however (and once we reject the entity orientation), there is no reason to separate objects into those which are intrinsically valued and those that are instrumentally valued. These distinctions are simply confused attempts to reify one element of countless, irreducible relational acts of individual valuers who experience value in many particular cultural or natural contexts. The belief that nature, or one of its elements, has intrinsic value is, according to this view, a confusion. A whole range of varied values must be recognized, respected, and reconciled, not sorted into one of two categories.

The question is not one of determining which objects have a reified value, but whether solid reasons can be given for invoking a particular value in a particular situation. This line of reasoning opens up the possibility of reconciling the intrinsic versus instrumental value debate, because it is possible to include both instrumental and noninstrumental reasons for preferring one set of policies to another without asserting that intrinsic value exists independently of human, valuing actions. If we reject this sharp dichotomy between instrumental and intrinsic value and the associated classification of natural objects as instruments or moral beings, a pluralist and integrative position emerges: there are many ways in which humans value nature and these ways run along a continuum, from entirely self-directed and consumptive uses (which includes spiritual and aesthetic value) to non-instrumental valuations. If a sharp distinction between these two opposed types of valuing is not made, the moral task of sorting entities into those that have and those that lack this special feature of non-instrumental value ceases to be a problem. The sorting question, that is, has interest only after one enters the bipolarized understanding of environmental values that comes with the web of assumptions shared by Economists and Deep Ecologists.

## 2 Rejecting the Entity Orientation

Suppose that we, following this more pluralistic approach, stop thinking of environmental evaluation as an exercise in categorizing objects at all. Rather, the goal will be to choose indicators of the adaptability of various technologies and policies. Our attention would then turn to the impact of existing and proposed technologies and policies on ecophysical and social processes, and the task would be to develop an indicator, or suite of indicators, that allow for the ranking of "development paths." A development path is a scenario that can be projected to unfold under a given policy or set of policies. The task of evaluation will then become one of ranking various present development processes that might unfold in the future. In the end, we want to be able to say that development Path A is more (less) likely to fulfill social values $V_1$, $V_2$, or $V_3$ than development Path B. This process approach simply ignores the problems and possibilities of reification and sets out to evaluate processes of development and change as they play out on a landscape.

Rejection of the entity bias has an even more profound implication for the theory of environmental value. If we reject the assumption that environmental evaluation is basically a matter of sorting entities and focus instead on evaluating processes, paths of change, and the values experienced by people and cultures within these processes, it is possible to recognize a deeper source of value in nature – what might be called nature's "creativity." Ilya Prigogine and Isabelle Stengers have argued persuasively that Western thought has for too long emphasized "being" at the expense of "becoming," and entities at the expense of processes.[34] Prigogine and other leaders of the emerging science of chaos and complexity have set out to repair this imbalance. They have argued that change, process, and becoming are more basic than being – that the world of objects we see is simply our stilted perception of a rich, multiscalar, evolving system.[35]

The more dynamic models suggested by the sciences of chaos and complexity place humans and their societies within a larger, evolving, ecological and physical system. All description and all evaluation occurs from within a dynamic system, so humans value nature from within nature. As parts of nature, humans are also conscious, autonomous agents who can, inadvertently or intentionally, affect the larger systems of which they are a part. Environmental values emerge from this key natural/cultural dialectic.

If this kind of thinking is applied to biodiversity policy, we can focus on the processes that have created and sustained the species and elements that currently exist, rather than on the species and elements themselves. Indeed, emphasis on the value of creative processes in nature may go a long way toward expressing the common factor in how many people value nature. When the Native animist worships or respects trees or animals, it is the trees' or animals' activity and presumed potency as well as their ability to affect human processes that excite religious impulses. When the agriculturalist or the forester values nature, it is the ongoing processes of productivity – the ability to provide a flow of useful products – that is the essence of nature's perceived value. Similarly, when the Deep Ecologist says that elements of nature have intrinsic or inherent value, this object-oriented statement is simply (in a more process-oriented vernacular) an insistence that there is majesty and meaning in the evolving processes of life. The common element of these different object-oriented statements of value is a correspondence between an important aspect of nature's ability to create and an important human impulse to value that creativity.

Similarly, it is reasonable to interpret advocates of biodiversity protection as valuing natural processes for their capacity to maintain, support, and repair damage to their parts. Once the categorical interpretation of questions of environmental valuation is avoided, instrumentalists and intrinsicalists may be able to adopt a common vocabulary and converging policies: that is, once the destructive implications of the entity orientation are exposed, the old dualism of utilitarian and intrinsicalism – the "either-or" mentality carried over from the instrumental/non-instrumental conceptualization – become "both-and" inclusions. It will then become time to move beyond the classification of objects to the giving of reasons to value, or abhor, certain trends and processes.[36]

The impulse to value nature's creativity – an impulse that, thankfully, has been expressed in many ways, by different persons and cultures – perhaps comes closer than either Deep Ecologists' or Economists' theories to capturing the deep and universal value that could unite all peoples behind an Earth Charter. Those theories are directed, one might say, at the specific content of people's values, rather than at the real and shared source of those specific values in nature. Emphasizing one value at the expense of others can only lead to conflict and division, because humans – struggling to survive in many local situations with differing constraints and opportunities and different natural and cultural histories – have different needs, preferences, and ideals. Some humans are hunters; some are birdwatchers; some are shamans; and others are developers and capitalists. When expressed in a dynamic process model, the common factor in all of these evaluative stances is the value of nature as a multi-scaled creative system.

This creativity is exhibited on many scales. In paleontology, it has resulted in great diversity; on the shortest temporal scale it gives hope of the next harvest to the faithful peasant who plants seeds. These creative processes, we can further say, are valued by humans because creative natural processes provide options and opportunities to fulfill human values, whatever those human values are. These values emerge from the human/nature dialectic of co-evolution; they do not exist in either the humans-only world of economists or in the independent realm of non-human values envisaged by deep ecologists. One way to articulate this viewpoint is to suggest that events take on meaning only within a cultural context and that environmental values must be woven into the experience and narratives of real people who live in actual cultures.[37]

An Earth Charter should not tell the many peoples and cultures of the world how they should value nature; rather, it should express the underlying value placed on nature's creativity and the opportunities to choose and to adapt that this creativity offers humans. This creative, sustaining force allows species to reproduce and maintain themselves and to create new, adaptive responses to changing eco-physical processes that form their environment. This value, it can be argued, exists at a deeper level than do the values of Economists and Deep Ecologists. Creativity – nature as a source – is the *sine qua non* for this value and other, more specific values.

This journey into foundational values has now taken us back to the the adaptive managers' position and their understanding of resilience. Resilience can be thought of as a promising attempt to identify and apply a characteristic of natural systems that is essential to the systems' continued creativity. A capacity to close the valuational loop has been added by our expansion of adaptive management to include a corrective to destructive human preferences within the management model. It is now possible to explain why individuals who value the future – those who are committed to living sustainably – *should* care about resilience. We value resilience because it allows a system to remain productive, maintain structure through energy dissipation, heal wounds, and repair stresses. These are the essential features of a system that maintains its creative force by maintaining its self-organizing, self-making structure.

Because it is the very basis of human opportunity, nature's creativity is valued both in the present and for the future. Making value analysis part of the policy process allows us to explain why individuals, who may value natural products for personal consumption, may also come to see how certain present consumptive patterns and the preferences that drive them, are inconsistent with maintaining opportunities and a range of free choices, opportunities to adapt, for future generations. Just as the smoker understands that continued smoking is inconsistent with good health, the car owner may someday realize that excessive use of fossil fuels is inconsistent with maintaining future opportunities. Driving a car is different than smoking, of course, since refraining from driving to protect future people includes an element of altruism, but both individuals must, ultimately, adjust their behaviour, and in doing so they will likely also alter their preferences. In both cases science provides data and models that alert consumers to a conflict between fulfilling one value and maintaining another. Social scientists should be able to model the resulting changes in preferences, after which cognitive psychology and related disciplines will become increasingly important aspects of community-based adaptive management.[38]

The more dynamic conception of values – those that would be appropriate within a process of adaptive management – emerges from the confluence of two initially separable but ultimately unifying forces. The first force comes from nature and must be understood by the natural sciences – it is the creative force that spun out our communities as a part of a far larger creation.[39] The second force is the striving of individual humans to survive, reproduce, and perpetuate their kind. This striving gives rise to many diverse goals in many different evolutionary contexts and to the ingenuity that is so valuable to humans. In this context, however, values should not be "attached" to objects, but understood as experiences that occur in countless real-life situations. The confluence of these intellectual forces thus affirms a crucial and deep connection between how humans value natural creativity, on the one hand, and human choice, freedom, and creativity on the other. The human choice and freedom that expresses itself in different values and in different persons and cultures' behaviour – from the hunter-gatherer's use value to the Deep Ecologists' inherent value – finds its ground in nature's creativity.

When we measure environmental values, therefore, we should also measure the extent to which the creativity of a natural system serves and is served by human creativity: that is, we should pay attention to how our actions and policies change these synergistic relationships. What is needed is a measure of how well a system maintains those forms of creativity that support a wide range of human opportunities, both in the present and in the future. Nature's creativity is experienced by us as a set of opportunities. The range of individual choices – an important prerequisite of human freedom – is thus affected by the range of nature's creativity. For example, the opportunities of a whittler or a wood sculptor are provided but also limited by the range of tree species at his or her disposal. Similarly, the opportunities – and range of choices – available to a housebuilder or developer is a reflection of the types and variety of landscapes and settings available. As discussed in Part 3, an OCI measure could capture these dynamic relationships.[40]

### 3 Avoiding Reductionism and Monism

One reason such sweeping consequences follow from the rejection of the entity orientation is that monism becomes irrelevant when people evaluate changes in the context of their culture. If we reject the monistic assumption, according to which all value must be explained according to a single principle, and no longer need to sort entities into those that are instrumentally and those that are inherently valued, it is possible to instead start from the pluralistic viewpoint, in which all cultures value nature and natural processes in their own ways.

As a first step, a vocabulary and operational measurements rich enough to express these multiple values must be developed. Pluralism as a working hypothesis can thus be embraced, as many values and types of values as possible can be characterized and operationalized. This leaves, for subsequent discussion, the question of whether some of these values can be usefully reduced to other values, assuming some level of multiple framework consolidation will eventually emerge. These evaluations are no longer constrained by the requirement that environmental values must be commensurable and measurable within a unified system of evaluation and with a single moral or evaluative "currency."

Monism and reductionism have been popular in environmental ethics and moral philosophy because a monistic system, which has only one principle to apply in all cases, avoids relativism or subjectivism. Monists reason that, if there are multiple principles, theories, and rules in the moral arena and some of these rules yield different conclusions in the same situation, moral evaluation will collapse into relativism.[41] While moral relativism should indeed be avoided, there are a number of strategies available for avoiding it within pluralistic systems – it is not necessary to embrace monism in order to avoid relativism. For example, it is possible to develop a two-tiered system of analysis in which the "action" tier includes multiple rules for choosing acceptable behaviour and a second, "meta" tier contains rational principles for deciding which of the action rules is appropriate in a certain situations.[42]

### 4 Rejecting the Placeless Evaluation Assumption

Evaluation models such as Economism and Deep Ecology are constrained by their monism to the expression of all value in a common currency. Their accounts of value thus tend to lose, in the process of aggregation, the place-relative knowledge and value that emerges within a specific dialectic between a human culture and its physical and ecological setting or context.[43]

One implication of the adaptational model for understanding environmental problems is an emphasis on the importance of localism. As we know from evolution, all adaptation takes place at the local level, as individuals "experiment" with various forms of adaptation to local conditions and either survive or fail to survive. As one relaxes the assumption that we need a single, universally aggregable accounting system for all environmental values, it becomes possible to hear, and register, the very real concerns of local cultures that are trapped between the hard realities of international economic forces beyond their control and the equally real limits and constraints that manifest themselves at the local and regional level. Localism (as a replacement for universalism) emphasizes local variation, diversity from locale to locale and from region to region, and many local "senses of place," each of which expresses a unique outcome at each particular place of the infinitely variable dialectics between local cultures and their habitats. Development and various development paths can therefore represent differing trajectories created by the nature/culture dialectic in a specific, culturally evolved place. This trajectory, given the above notions, can be measured on an economic scale and by using an OCI that measures the extent to which the development trajectory maintains and increases ecophysically based opportunities, using the multi-generational scale required to judge a culture's true sustainability.

CONCLUSION

The attempt to develop an Earth Charter, a document that expresses "the shared values of people of all races, cultures, and religions," may provide an opportunity to re-examine current approaches to environmental valuation and provide clarification of issues that form the essential cultural backdrop for developing an ethic governing the practice of conservation biology.[44] This chapter has argued that neither the narrow, human-centred utilitarianism of Economism nor the assertion of Deep Ecology can characterize a universal value that will unite humankind behind an effort to protect biological diversity. Both of these widely espoused, yet opposing theories share a set of assumptions about the nature of environmental value – the importance of monism, a sharp separation between intrinsic and instrumental values, and an object orientation. These assumptions are also, when examined objectively, highly vulnerable. The discussion of environmental values within such practical contexts as the discussion of conservation priorities is usually entity-orientated and sees problems of conservation as problems of sorting objects and saving inventories. As environmental management everywhere moves toward more holistic and process-oriented management models, entity orientation seems increasingly obsolete.

The adaptive management model emerges instead as a way of understanding human-nature interactions from the viewpoint of a community adapting to a larger, changing ecophysical system. This approach provides a simple, plausible model of how environmental problems emerge at the juncture of human, technological, and environmental change. Collective individual choices, in response to the opportunities and constraints offered by the environment of individuals in the first generation, can change the environment so that individuals of the second generation face a much poorer range of options. Environmental problems emerge as cross-generational spill-over effects, effects that reduce the range of choices that will be faced in the future.

Thus, while everyone must derive economic sustenance from their environment, a concern for the future demands that we also monitor the impacts of current actions on our mix of opportunities and constraints. It is therefore reasonable for members of freedom-loving human communities to maintain a non-declining set of opportunities, based on possible future uses of the environment. This responsibility is based on a sense of community with peoples of the future and on a sense of fairness: those who will live in the future ought not to face, as a result of our actions today, a seriously reduced range of options and choices as they try to adapt to the environment. Acceptance of this responsibility as an important aspect of an adaptive management model, however, requires that the model allow the reconsideration of values and preferences when evidence suggests that current values and behaviour are likely to reduce the amount of opportunities, and increase the amount of constraints, that will be faced.

Building on this adaptational model (and avoiding the reduction of the many values humans derive from nature to a single type), it is possible to see that what is valued in common by persons with diverse relationships to nature is its creativity. Nature's creativity provides options that are the basis for human opportunity. This is a level of environmental value that may well be universal. While the hunter, the developer, the shaman, and the birdwatcher all exercise very different individual values and options – some self-oriented and some not – what they all share is a powerful dependence on the creative aspect of nature. If, then, we can avoid the assumptions – of monism, of a sharp dichotomy between intrinsic and instrumental value, and of the entity orientation – that bind deep ecologists and economists in polarized opposition, it may be possible to find, in a celebration of nature's infinite creativity, a universal value capable of supporting a truly unifying Earth Charter.

If this universal value were embraced, the goals and the roles of conservation biologists would become clearer and more manageable. As noted in Part 1, conservation biologists often work in areas of the world where cultures and values differ greatly from their own. The values outlined here suggest a better way of understanding the interactions of conservation biologists, who value nature in their own ways, with members of other cultures, who value nature in quite different ways. Individuals who encounter nature in many different natural and cultural contexts should be expected to express diverse valuations in diverse ways. This diversity should be understood as the valued expression of countless unique and interesting dialectics between culture and nature. When this diversity leads to distrust and cross-cultural conflict, however, it should also be remembered that the diversity of human valuations merely reflect the universal but infinitely varied interdependence of all human cultures on the unquestionable value of nature's creativity as a necessary context for human creativity and diversity.

### NOTES

1 This effort is being organized by the Earth Council and Green Cross International with support from the Dutch government. For more information, see the special issue of *Earth Ethics* 7 (Spring/Summer 1996), nos. 3 and 4. For more, and more recent, information, see www.earthcharter.org

2 Steven C. Rockefeller, "Global Ethics, International Law, and the Earth Charter," Earth Ethics, 3–5.

3 Ibid.

4 These terms are used in this chapter as labels for advocates of each theory of value. It should not be inferred, of course, that all economists fit the label "Economist," as defined and used here; similarly, the phrase "Deep Ecologist" refers more to a caricature based on the characteristic belief of these practitioners that nature has value independent of humans and their motives. Since the purpose of this paper is to examine the larger picture, details and possible differences within these categories are not that important. For a more detailed explanation of Economism, see William Baxter, "People or Penguins," and A. M. Freeman, "The Ethical Basis of the Economic View of the Environment," in D. VanDeVeer and C. Pierce, *The Environmental Ethics and Policy Book* (Belmont, CA: Wadsworth Publishing Company, 1994). For an explanation of Deep Ecologism see, for example, Holmes Rolston, III, *Conserving Natural Value* (New York: Columbia University Press, 1994), especially chapter 6.

5 See, for example, "The Earth Charter (Working Draft for Benchmark Draft II)," available from Steven Rockefeller, Chair, Earth Charter Drafting Committee, Middlebury, Vermont.

6  Gifford Pinchot, *Breaking New Ground* (Covelo, CA: Island Press, 1987), 323.

7  Heraclitus, it appears, was the first "new ecologist," because he empha-sized the importance of change in nature. See Bryan G. Norton, "Change, Constancy, and Creativity: The New Ecology and Some Old Problems," in *Duke Environmental Law and Policy Forum*, 7 (Fall, 1996): 49–70.

8  See John Dewey, "The Influence of Darwinism on Philosophy," in John Dewey, *The Influence of Darwin on Philosophy, and Other Essays in Contemporary Thought* (New York: Henry Holt and Company Inc., 1910).

9  See Stuart L. Pimm, *The Balance of Nature?: Ecological Issues in the Conservation of Species and Communities* (Chicago: University of Chicago Press, 1991); and Bryan G. Norton, "A New Paradigm for Envi-ronmental Management," in R. Costanza, B. Norton, and B. Haskell, eds., *Ecosystem Health: New Goals for Environmental Management* (Covelo, CA: Island Press, 1992).

10  Christopher Stone, *Earth and Other Ethics* (New York: Harper and Row Publishers, 1988), 116.

11  See Bryan G. Norton, *Why Preserve Natural Variety?* (Princeton, NJ: Princeton University Press, 1987). Chapters 12 and 13 provide a history and criticism of attempts to "prioritize" taxonomic categories as a basis for setting policy in the United States Fish and Wildlife Service.

12  William Cronon, *Changes in the Land* (New York: Hill and Wang, 1983).

13  The three phases should be understood as policy phases: individual scien-tists and many groups of scientists developed alternative formulations of the underlying biological relationships, but the focus of this discussion is the formulation of biological protection issues in policy and legislative contexts.

14  See Bryan G. Norton, "Biological Resources and Endangered Species History, Values, and Policy" in Jeffrey McNeely and Lakshman Guruswamy, *Protection of Biodiversity: Converging Strategies* (Durham, NC: Duke University Press, 1998) for a more detailed discussion of the value and implications of the Endangered Species Act.

15  The symposium was held in Washington, DC, in September 1986. See E.O. Wilson, ed., *Biodiversity* (National Academy Press: Washington, DC, 1988) for proceedings of the forum.

16  While this simple, stark definition does not appear explicitly in Wilson, it served as a working definition for the symposium. For an equivalent defi-nition, see B.A. Wilcox, "In Situ Conservation of Genetic Resources: Determinants of Minimum Area Requirements," In J.A. McNeeley and K.R. Miller, eds., *National Parks, Conservation, and Development: The Role of Protected Areas in Sustaining Society*, Proceedings of the World

Congress on National Parks, Bali, Indonesia, October 1982 (Smithsonian Institution Press: Washington, DC, 1984).

17 See, for example, R. Costanza, Bryan G. Norton, and B. Haskell, *Ecosystem Health* (Covelo, CA: Island Press, 1992).

18 The issue of inherent value emerges again in this context as some authors, such as J. Baird Callicott (*In Defense of the Land Ethic* (Albany, NY: State University of New York Press, 1989)) and Laura Westra (*An Environmental Proposal for Ethics: The Principle of Integrity* (Lanham, MD: Rowman and Littlefield Publishers, 1994)) argue that the term "integrity" signals an acceptance of some form of nonanthropocentric evaluation. Other authors, including most of them represented in *Ecosystem Health*, use the terms analogically, and draw no ontological or deontological conclusions from attributing "health" or "integrity" to ecosystems.

19 I have discussed these linguistic aspects in more detail in "Improving Ecological Communication," *Ecological Applications*, 8 (1998): 350–64.

20 See Kai Lee, *Compass and Gyroscope* (Covelo, CA: Island Press, 1993) and L.H. Gunderson, C.S. Holling, and S.S. Light, eds., *Barriers and Bridges* (New York: Columbia University Press, 1995).

21 See Michael Jacobs, "What Is Socio-Ecological Economics?" *Ecological Economics Bulletin* 1 (April 1996); and, "Reflections on the Discourse of Politics of Sustainable Development, Part I," Manuscript, February 1995; and *The Green Economy* (London: The Pluto Press, 1991).

22 See R. Therivel, E. Wilson, S. Thompson, D. Heaney, and D. Pritchard, *Strategic Environmental Assessment* (London: Earthscan, 1992); R. Therivel and M.R. Partidario, eds., *The Practice of Strategic Environmental Assessment* (London: Earthscan, 1996); and, for a review and application to urban planning, see A. Shepherd and L. Ortolano, "Strategic Environmental Assessment for Sustainable Urban Development," *Environmental Impact Assessment Review*, 16 (1997): 331–5.

23 See, for example, Sym Van Der Ryn and Stuart Cowan, "Nature's Geometry," *Whole Earth Review* (Fall 1995): 9–14; Arie Rip, "Controversies as Informal Technology Assessment," *Knowledge Creation, Diffusion, Utilization*, 8(1986): 349–71.

24 C.S. Holling, *Adaptive Environmental Assessment and Management* (London: John Wiley & Sons, Inc., 1977); C.J. Walters, *Adaptive Management of Natural Resources* (New York: MacMillan Publishing Co., 1986); K. Lee, *Compass and Gyroscope* (Covelo, CA: Island Press, 1993); and Gunderson, Holling, and Light.

25 Mick Common and Charles Perrings, "Towards an Ecological Economics of Sustainability," *Ecological Economics* 6 (1992): 7-34; and C.S. Holling, "Engineering Resilience versus Ecological Resilience," in P.C. Schulze, ed., *Engineering within Ecological Constraints* (Washington, DC: National Academy Press, 1996).

26 See H. A. Regier and James J. Kay, "An Heuristic Model of Transformations of the Aquatic Ecosystems of the Great Lakes–St. Lawrence River Basin," *Journal of Aquatic Ecosystem Health* 5(1996): 3–21, for further discussion of "flips."

27 L.H. Gunderson, C.S. Holling, and S.S. Light, "Barriers Broken and Bridges Built: A Synthesis" in Gunderson, Holling, and Light.

28 See H. Daly and J. Cobb, *For the Common Good* (Boston: Beacon Press, 1989); and R. Costanza, ed., *Ecological Economics: The Science and Management of Sustainability* (New York: Columbia University Press, 1991).

29 See Bryan G. Norton, "Evaluating Ecosystem States: Two Competing Paradigms" *Ecological Economics* 14 (1995): 113–27, for a more detailed analysis of this debate.

30 K. Arrow, B. Bolin, R. Costanza, P. Dasgupta, C. Folke, C.S. Holling, B.-O. Jansson, S. Levin, K.-G Maler, C. Perrings, and D. Pimentel, "Economic Growth, Carrying Capacity, and the Environment," *Science* 268: 520–1.

31 Lee, *Compass and Gyroscope.*

32 Ibid., 192.

33 J. Baird Callicott, *In Defense of the Land Ethic* (Albany, NY: State University of New York Press, 1989).

34 I. Prigogine and I. Stengers, *Order Out of Chaos: Man's New Dialogue with Nature* (New York: Bantam Books, 1984).

35 Leading scientists have thus joined Heraclitus, Henri Bergson, and A.N. Whitehead in embracing a process-oriented understanding of natural systems. See Henri Bergson, *Creative Evolution* (New York: Holt, Rinehart & Winston, 1911), and A.N. Whitehead, *Process and Reality* (New York: The Macmillan Co., 1929). Space does not permit a full discussion of these topics. The reader is referred to any one of a number of popular and semi-popular discussions of complexity and chaos.

36 My position is similar and owes much to that of Paul Wood in "Biodiversity as the Source of Biological Resources: A New Look at Biodiversity Values," *Environmental Values* 6 (1997): 251–68.

37 As has been suggested to me by Ken Cussen, in private correspondence and in exchanges of unpublished manuscripts.

38 See Bryan G. Norton, "Thoreau's Insect Analogies: Or, Why Environmentalists Hate Mainstream Economists," *Environmental Ethics*, 13 (1991): 235–51; B. Norton, "Economists' Preferences and the Preferences of Economists," *Environmental Values* 3 (1994): 311–32; and B. Norton, R. Costanza, and R. Bishop, "The Evolution of Preferences: Why 'Sovereign' Preferences May Not Lead to Sustainable Policies and What to Do about It," *Ecological Economics* 24 (1998): 193–212, for further discussion of how preferences change across time and in response to information on the impact of current preferences and actions.

39 There is no intent to suggest that the "creative force" mentioned here involves *intentional* creation. Unfortunately, all terms and phrases that capture the creative aspect of natural systems seem also to imply intentional creation. This is a defect in our (unfortunately) dualistic everyday language. See Prigogine and Stengers for an explanation of these points.

40 While space does not permit a full development of these ideas, I have examined the intellectual and methodological advantages that emerge when the entity orientation is abandoned. See also Bryan G. Norton, "Ecological Integrity and Social Values: At What Scale?" *Ecosystem Health* 1 (1995): 228–41.

41 See, for example, J. Baird Callicott, "The Case against Moral Pluralism," *Environmental Ethics* (12) 1990: 9-24.

42 See Bryan G. Norton, *Toward Unity Among Environmentalists* (New York: Oxford University Press, 1991), 200, for an outline of such a system. Also see Costanza, Norton, and Bishop for an explanation of how a two-tiered system might work in practice.

43 See Bryan G. Norton and B. Hannon, "Environmental Values: A Place-Based Theory," *Environmental Ethics* 19 (1997): 227–45; and B. Norton and B. Hannon, "Democracy and Sense of Place Values," in A. Light and J.M. Smith, *Philosophy and Geography, III*: Philosophics of Place (Lanham, MD: Rowman and Littlefield, 1998).

44 Rockefeller, "Global Ethics, International Law, and the Earth Charter," 3.

7 The Notion of Effectiveness
*Lessons from the Field of*
*International Development*

GEORGINA WIGLEY AND
HEATHER BASER

EDITORS' INTRODUCTION

Several chapters of this book have stressed that cultural sensitivity is essential to making conservation actions efficient. Unfortunately, the training provided to conservation biologists in most institutions of higher learning focuses on understanding plants, animals, and ecology, and rarely includes an introduction to relevant social sciences, such as sociology, anthropology, and psychology. While the implications of cultural diversity and the need for cultural sensitivity are fairly new concepts for biologists, they have been central to the thinking of students and practitioners of international development for a few decades. In this chapter, Georgina Wigley and Heather Baser of the Canadian International Development Agency (CIDA) review some tools that were developed to identify talents and personality traits in candidates for overseas work. These talents are important to the establishment of positive, fruitful intercultural relationships. They believe that technical competence alone is not enough for efficient action: an understanding of human psychology and an acceptance of differences are also essential elements of international collaboration.

INTRODUCTION

Every year, tens of thousands of Northern nationals take up assignments in developing countries as part of official development assistance or development co-operation programs. These projects fall into a general area of activity known as technical co-operation, the objective of which is to assist developing countries to find and implement their own long-term solutions to development problems through improving the level of skills, knowledge, know-how, and productive aptitudes in their populations.[1] Long-term expatriate advisors (on assignments of several years) or short-term consultants (on assignments of, usually, fewer than six months) work with Southern institutions on many levels or help develop training programs for Southern nationals. Such projects can also provide funding for research programs, seminars, and equipment.

Development agencies are, understandably, concerned that the technical co-operation programs they sponsor are as effective as possible and that they make a successful contribution to the fostering of sustainable development in recipient countries. Many of the lessons learned in the field of international development about what influences success in technical co-operation apply equally to other fields, such as conservation biology, a field in which interaction between Northern and Southern partners is common.

Two key elements underpin the success of technical co-operation programs. These elements are also relevant to conservation biology. First, the programs and mandates of advisors assigned to a country must be well designed so that they actually address the problems that limit sustainable development in that country. Second, expatriate advisors must be successful in what they are sent overseas to accomplish. As donors and other international agencies study which factors contribute to success or effectiveness in these two areas, it is becoming increasingly evident that non-technical considerations, such as development ethics and the behaviour and attitudes of expatriate advisors in host-country societies, play an important role. This chapter reviews some of the work now being done, in order to demonstrate the importance of ethical consideration in the design of technical co-operation programs. It then considers behavioural and attitudinal criteria in the selection of advisors assigned to these programs. Finally, it discusses lessons that are applicable to conservation biology.

## THE EFFECTIVENESS OF TECHNICAL
## CO-OPERATION: WORK IN PROGRESS

The debate that focuses on the effectiveness of technical co-operation includes a consideration of development ethics, "the ethical reflection on the ends and means of global development that lay bare the value questions posed by development theory, planning and practice."[2] At issue are fundamental questions about the goals of development and its moral consequences. It is increasingly being recognized that sustainable development, and the development of local capacities that underlies it, are endogenous processes in which outsiders can only give support or act as catalysts. Hence, key questions that must be asked when designing development programs include What is this development for and by whom will it be carried out?, and What means are acceptable to reach the stated ends, and Who decides on these?

As has been made clear by Lala H. Rakotovao et al. (see chapter 5), and Rogelio Cansarí (see chapter 3), similar types of ethical questions should be asked by conservation biologists planning research or other projects in Southern countries. For example, a particular element of the natural environment may have a different meaning or value to a person living in a Southern country than it does to a person from a Northern country (see chapter 4), and incorrect assumptions about such meanings may lead outsiders to take inappropriate decisions or actions.

As discussed by Morgan and Baser, more thoughtful consideration of these non-technical questions by people in the development community is gradually leading to changes in the design of technical co-operation programs.[3] In 1991, the Development Assistance Committee (DAC) of the Organization for Economic Co-operation and Development (OECD), a donor-nation policy forum, issued the guideline *Principles for New Orientations in Technical Co-operation*. This document suggests how to avoid all-too-common pitfalls when designing and implementing technical assistance programs. Problem areas for many development programs include an over-reliance on expensive and temporary expertise from Northern countries at the expense of nurturing competent local expertise and undermining of sustainable initiatives by taking a piecemeal, project-by-project approach that is driven more by donors' agendas than by Southern countries' needs and priorities. These lessons apply equally to the design of conservation-biology projects.

Donor agencies have responded to challenges such as those outlined in the OECD guideline by sponsoring research into defining and measuring effectiveness as well as by re-evaluating their policies, practices, and approaches to design-development programs that explicitly take into account social and cultural factors. To make programs more demand-driven (and less donor-driven) an increased effort is being made to involve recipient-country representatives in program design and expatriate personnel selection. Donors are also increasingly rethinking their approaches and becoming more inclined to use techniques, such as capacity development, that change the nature of donor involvement. Capacity development emphasizes building on local interests and expertise in the conception and design of projects and involving those who are interested in and/or affected by an initiative in all aspects of its conception, planning, implementation, and evaluation.

The urgent need for increased capacity building has been previously stressed by Léonard Mubalama, (see chapter 1). Participatory approaches, however, will require adjustments to the traditional "expert-apprentice" approach often taken in the design of technical co-operation projects, in which the foreign researcher is the "expert" and the national researcher is seen as subordinate, with little knowledge or expertise to contribute. Because of this assumption, the foreign researcher typically takes charge, which reduces the national researcher's sense of responsibility for results produced and limits his or her opportunity to learn through experience.

These new approaches will also require more emphasis on recruiting advisors with strong interpersonal skills and of appropriate attitudes and behaviour (as discussed in the next section). As Jean Bossuyt writes, "If capacity development is about helping people to help themselves, then donors need to move away from relationships characterized by prescription, imposition, condition-setting and decision-making, to ones characterized by explanation, demonstration, facilitation and advice. If [developing countries' inherent strengths] are to be nurtured, donor agencies need to revise how they deliver external inputs."[4]

Although it is not easy to effect change in the face of entrenched attitudes and procedures that characterize many development institutions, a lively debate is underway, and new avenues for technical co-operation are being explored.

## CHARACTERISTICS OF EFFECTIVE TECHNICAL ADVISORS

While fundamental ethical questions must certainly be debated and open-minded reforms to program design attempted if technical co-operation is to become more effective, the question of individual advisor effectiveness is a crucial element in the success of technical co-operation. Advisor effectiveness is related to an individual's ability to exchange or transfer experience in a given technical area. Research into individual effectiveness is demonstrating that while technical knowledge and experience are essential qualifications, other non-technical considerations, linked to core traits that shape the advisor's behaviour in a cross-cultural context or environment, play an equally if not more important role. Canada and the Nordic countries have been especially active in uncovering what the characteristics of effective technical advisors are and, subsequently, identifying and selecting those who have such characteristics to fill advisor positions.

Early work (in 1979) by Brent Ruben and Daniel Kealey found that patterns of success in overseas adaptation could be predicted by seven interpersonal and communication skills: (1) empathy; (2) respect; (3) role behaviour; (4) non-judgmentalness; (5) openness; (6) tolerance for ambiguity; and (7) interaction management.[5] "Canadians in Development",[6] a follow-up study by Kealey and Frank Hawes, reiterated the importance of these seven skills but also found that only a small percentage of Canadians serving abroad were effective in transferring their experience and skills to their overseas assignments. Inefficiency occurred because many of these Canadians were unsuccessful in dealing with the counterparts with whom they were assigned to work in the Southern country. Another study confirmed a similar phenomenon among Nordic technical advisors: only about 20 per cent were judged highly effective in their ability to transfer skills and knowledge.[7]

In the late 1980s, Daniel Kealey conducted over a thousand interviews with a variety of people involved in overseas assignments. He obtained their opinions on the types of attitudes and

Criteria for Effectiveness Most Frequently Cited by Representatives of Different Sectors
Involved in Work Overseas

| Sector | Criteria mentioned most frequently by representatives of the sector | | |
|---|---|---|---|
| | First | Second | Third |
| Donors | Flexibility | Initiative | Motivation |
| Technical advisors in the field | Tolerance | Patience | Marital/family stability |
| Recipients of technical co-operation | Technical expertise | Understanding and acceptance of local culture | Social participation skills |
| Corporate managers | Technical expertise | Management ability | Previous overseas experience |
| Field Managers (private sector) | Spouse's attitude | Understanding of local culture | Tolerance |
| Voluntary organizations | Motivation | Interpersonal skills | Flexibility |

Source: D.J. Kealey, *Profile of Technical Advisors for International Development* (report prepared for
the Canadian International Development Agency, 1994).

skills that they felt were most likely to contribute to an individual's
success overseas. Although the interviewees represented a wide
range of sectors (donor agencies, recipient countries, advisors in
the field, organizations that send volunteers overseas, and private
companies that hire expatriate advisors), there was remarkable
consensus on the importance of non-technical skills to overseas
effectiveness. Apart from technical expertise, the following criteria
were identified as important by all the interviewees: (1) tolerance;
(2) patience; (3) respect; (4) empathy; (5) flexibility; (6) listening
ability; (7) interest in the local culture; (8) cross-cultural sensitivity; (9) communication skills; (10) commitment; and (11) realistic
expectations.

These findings led to further research, sponsored by CIDA, to
develop assessment instruments or techniques to help identify
whether candidates for advisor jobs have the traits associated with
success and to recommend what can be done to improve performance in weak areas. The research produced four types of instruments that can be used to help identify the most appropriate individuals to fill overseas positions: (1) model technical advisor

profiles; (2) a "self-assessment" questionnaire, to be filled in by candidates; (3) cross-cultural case studies; and (4) a reference rating form. It also developed an integrated process for the recruitment, screening, selection, and preparation of technical advisors. These instruments are discussed in detail below. Conservation biologists and others who are interested in ensuring that they are as prepared as possible for successful overseas work may find these instruments useful.

## 1 Model Technical-Advisor Profiles

The generic model technical-advisor profile describes the skills or traits associated with success overseas. It focuses on three non-technical skills: adaptational skills; cross-cultural skills; and partnership skills. It also highlights the types of knowledge and experience that seem to contribute to cross-cultural effectiveness.

Adaptation skills are those skills that help a person with living and working conditions overseas. Strong adaptational skills help people develop a sense of well-being and comfort, and to feel that they are "at home" in the host culture. They also help an individual to react appropriately when faced with an unexpected or unfamiliar situation. These skills include positive attitudes, flexibility, patience, emotional maturity, and inner security.

Cross-cultural skills help a person participate in a local culture and find culturally appropriate ways of living and working within that culture. Key traits that contribute to cross-cultural effectiveness include realism (having a realistic sense of the working and living conditions that one is likely to encounter in the field), tolerance, the interest and desire to become involved in other cultures, cultural sensitivity, and political astuteness.

Partnership skills help workers cope with the professional demands of the assignment and the need to establish effective working relationships with national colleagues and partners. As Léonard Mubalama's case study illustrates (see chapter 1), traditional approaches of "apprenticeship," in which the expatriate researcher is always the expert and the national researcher is always an apprentice, are not only ineffective but also offensive and neo-colonialist. A better approach is to work as colleagues, so that both individuals learn from and support each other in ways that lead to a two-way transfer and acquisition of skills. Core skills necessary for this kind of partnership building include a mix of personal skills and qualities, such as openness to others, professional commitment, perseverance, initiative, a desire to build relationships, and self-confidence.

In addition to the generic model technical advisor profile, specific profiles for three categories of technical advisor were also developed: team leader; long term advisor; and short term advisor.

## 2  Self-assessment questionnaire

A second tool developed by CIDA to identify appropriate candidates for overseas positions was a questionnaire. Candidates answered questions about their knowledge, expectations, and attitudes. This self-assessment instrument provided a supplement to assessment instruments in which candidates are judged by others. This questionnaire, the Overseas Effectiveness Inventory, includes 119 questions and takes approximately thirty minutes to complete. The test items are based on the characteristics associated with effective technical advisor performance outlined in the model advisor profiles,[9] including (1) relevant background information; (2) knowledge of and attitudes toward development; (3) expectations about the assignment and the country of posting; (4) attitudes toward living and working overseas; (5) attitudes toward learning a new language; (6) understanding of their role overseas; (7) motivations for wanting the assignment; (8) interpersonal skills; (9) management/work style; (10) achievement orientation; (11) attitudes of spouse and children; (12) need for security; (13) need for social status; and (14) task vs. process orientation.

Ideally, the responses are analyzed by a psychologist trained to assess the questionnaire. This analysis then forms part of the information used by a selection committee to identify individuals suitable for various overseas assignments.

## 3  Cross-cultural Case Studies

The third step in the selection process looks at a series of seven case studies that are based on situations an advisor in a given job category may face during an overseas assignment. These case studies include the following situations:

1 Organizational restructuring (examines the challenge of a six-month assignment);
2 The coming of a new team leader (examines the difficulties of managing a team: for example, who should be the team leader – an expatriate researcher or the national researcher?);
3 The challenge of making connections (examines intercultural communications problems);
4 A needs assessment consultation (examines the challenge of a three-month assignment);
5 An advisor's reflections (examines how an individual assesses his or her behaviour during an assignment);
6 The selection of an advisor (examines the task of screening technical co-operation personnel);
7 The selection of an advisor (again, examines the task of screening technical co-operation personnel).

The responses to the open-ended questions included in the case studies provide the selection committee with insights into the ability of candidates to deal with typical personal and professional situations. As with the Overseas Effectiveness Inventory, the responses to the case studies are ideally assessed by a psychologist who is associated with the selection committee and who is trained in the use of these instruments.

*4 Reference Rating Form*

Previous research has established that when peers and supervisors rate candidates systematically and cover the factors critical to overseas effectiveness, they are powerful predictors of performance.[10] The Interpersonal Behavioural Assessment Index asks a candidate's referees to rate that candidate's behaviour in social and professional settings using criteria from the Overseas Effectiveness Inventory. The index is divided into three sections:

1 An assessment of the candidate's level of competence and suitability, taking into consideration relationship building, respect, openness, marital and family stability, and professional style;
2 A rating of the degree to which different statements describe the behaviour of the candidate; and

3 An assessment of the candidate's behaviour when dealing with five different work situations. For example, if the candidate was assigned to a managerial position, how would he or she be likely to respond in the midst of staff conflicts, when holding his or her first staff meeting, or when dealing with a difficult staff member? The possible answers from which referees may choose are designed to provide an indication of the referee's perception of the candidate's skill levels in areas such as interpersonal effectiveness and cross-cultural sensitivity and his or her level of self-reliance and self-confidence.

### 5 Developing an Integrated Process for Recruitment, Screening, Selection, and Preparation of Technical Advisors

While the research discussed above focuses on the development of methods to screen and select candidates, the researchers recognize that even good selection instruments cannot in themselves guarantee that technical advisors will adapt and perform effectively on an overseas assignment. They understand that it is also important to correctly scope and design the assignment and to prepare an accurate job description. These actions help identify the characteristics required of the advisor who will fill the position. Another crucial step, to be carried out after a well-qualified candidate has been identified, is to link the results of a candidate's selection process with the design of an appropriate training package, in order to address areas of weakness that were identified through the use of the selection tools.

In addition to advocating an integrated system that supports the selection and training of new advisors, researchers also want the tools and overall process that they are developing to be used by development agencies and others who are involved in the delivery of technical assistance programs. These tools and processes may also be adapted by people in other fields (such as conservation biology) who work overseas.

### CONCLUSION: LESSONS FOR CONSERVATION BIOLOGISTS

This chapter has examined factors that influence the effectiveness of technical co-operation in developing countries. At the level of

Main Components of an Integrated System Designed to Enhance Technical Advisor Effectiveness

|  | Project Scoping Phase | Advisor Recruitment and Selection | Advisor Training and Preparation |
|---|---|---|---|
| TASKS | 1 Job analysis and training needs | Pre-screening program (self-assessment) | Design based on assessment |
|  | 2 Socio-cultural analysis | Selection committee identification | Establishment of training contract with individual advisors |
|  | 3 Institutional analysis | Selection instrument administration | Training |
|  | 4 Counterpart screening | Selection interviews | Overseas performance monitoring |
| OUTPUTS | 1 Project objectives identified | Advisors selected with best potential to succeed | Trained advisors equipped with knowledge and skills needed to succeed overseas |
|  | 2 Scope of work including job requirements and constraints identified | Training needs of individual advisors identified | N/A |
|  | 3 Selection criteria for technical advisors established | N/A | N/A |

Source: D.J. Kealey, Profile of Technical Advisors for International Development (report prepared for the Canadian International Development Agency, 1994).

program design, it is becoming increasingly clear that ethical considerations are an important element in the effectiveness of technical co-operation programs and projects. Program design must also take into account the social, cultural, and institutional context of the host country. Ensuring that host-country counterparts participate in the design is one of the best ways to achieve this objective. As well, initiatives must be designed with sustainability in mind: ensuring that local capacity exists or is developed to sustain the initiative's achievements after the external support is phased out.

This chapter has also looked at the research into what traits or behavioural characteristics effective technical co-operation personnel share and how these traits can be assessed in potential candidates. Research has demonstrated that effective performance in a different culture requires much more than that a person be professionally knowledgeable or competent. A host of non-technical factors also shape people's behaviour in a cross-cultural environment, and hence affect their ability to complete what they have been sent overseas to do.

These findings are supported by the personal experiences of L.H. Rakotovao, Léonard Mubalama, and Rogelio Cansarí (see chapters 5, 1, and 3, respectively). These authors raise concerns about the negative impact of foreign scientists' inappropriate behaviour on host-country nationals and local communities. By eroding confidence and credibility and generating resentment and, in the worst cases, even hostility, insensitive foreign researchers can impair the ability of other expatriates to carry out effective research and conservation work overseas.

For these reasons, the research into effectiveness discussed in this chapter can offer lessons to conservation biologists and others who find that their work takes them to Southern countries. At a minimum, researchers should ensure, as much as possible, that projects they design take into account ethical questions, such as the possible impact on traditional values and the cultural identity of people in the target country. Those people who work overseas should also pay special attention to their non-technical abilities. They should seek training or coaching in areas of weakness or read instructional material that can help them develop skills in these areas. A person who is most likely to be effective in a cross-cultural environment is one who has strong interpersonal and communications skills as well as competence in a scientific or technical area.

Key qualities that candidates for overseas work should posses include: facilitation; a willingness to listen and learn from others; sensitivity to cross-cultural issues; empathy (being able to see many sides of an issue); the ability to understand the context in which development takes place; and the ability to take a broad, systems approach to problem solving. While working in the field, an effective person promotes participation and is a catalyst for learning and action by others; emphasizes partnership and sharing control; is discrete, is respectful of different cultures' values, beliefs, schedules, and ways and is flexible in adapting to these where appropriate;

and is sensitive to the choice of technologies and promotes those that are most appropriate.

Finally, a person who is most effective overseas is very conscious of the assumptions under which he or she operates and understands that assumptions that were valid at home will not necessarily be valid in another culture. This point has been made in chapter 4 by Priscilla Weeks et al. A cross-culturally effective person regularly examines his or her assumptions by listening to the views and opinions of host country nationals at all levels, and makes an effort to adopt country-specific approaches to work and social interaction whenever possible.

## NOTES

1  Organization for Economic Co-operation and Development (OECD), *Principles for New Orientations in Technical Cooperation* (Paris: Organization for Economic Co-operation and Development, 1991).

2  International Development Information Centre (IDIC), *Development Ethics*, Development Express 2 (1996), 3, (based on work by Glenda Robertson, consultant).

3  P. Morgan and H. Baser, *Making Technical Cooperation More Effective: New Approaches by the International Development Community* (Ottawa: Technical Cooperation Directorate, Canadian International Development Agency, 1993).

4  J. Bossuyt, *Capacity Development: How Can Donors Do It Better?* Policy Management Brief no. 5 (Maastricht: European Centre for Development Policy Management, 1995), 1.

5  D.J. Kealey and B. Ruben, "Cross-Cultural Selection Criteria, Issues, and Methods," in D. Klandis and R. Brislin, eds., *Handbook of Intercultural Training* (New York: Pergamon, 1983).

6  D.J. Kealey and F. Hawes, *Canadians in Development: An Empirical Study of Adaptation and Effectiveness on Overseas Assignment* (a technical report prepared for the Canadian International Development Agency, 1979).

7  K. Forss, J. Carlsen, E. Froyland, T. Sitari, and K. Vilby, Evaluation of the Effectiveness of Technical Assistance Personnel (report to DANIDA (Danish International Development Assistance), FINNIDA (Finnish Ministry of Foreign Affairs, Department for International Development Cooperation), NORAD (Norwegian Development Cooperation), and SIDA (Swedish International Development Cooperation Agency), 1988).

8  D.J. Kealey, *Profile of Technical Advisors for International Development* (report prepared for the Canadian International Development Agency, 1994).

9  D.J. Kealey, *Overseas Screening and Selection Project: Final Report* (Technical Cooperation Directorate, Canadian International Development Agency, 1994).

10  Ibid.

11  Kealey, *Profile of Technical Advisors for International Development.*

# 8 Conservation That Makes Dollars and Sense
## *The* RARE *Center for Tropical Conservation Work in the Caribbean*

PAUL BUTLER AND
VICTOR ALLEYNE REGIS

### EDITORS' INTRODUCTION

When Victor Alleyne Regis of RARE Center for Tropical Conservation presented the following paper at the 1996 World Conservation Congress, many attendants were very excited by what he had to say. His energy and enthusiasm put their understanding of the positive contribution that conservation efforts can make in a whole new light. This summary of RARE Center philosophy, initiatives, and accomplishments provides an excellent concluding chapter for this book. RARE Center "helps people help themselves," and its actions rely on the dedicated involvement of local communities and individuals. We hope that all conservation biologists who strive to protect biological diversity in developing countries will share our reading of this chapter – that it is by relying on local people, not by viewing them as opponents, that we will be successful.

### A HOLISTIC, LOCALLY BASED APPROACH

Whether it's St. Lucia's rugged volcanic beauty or the pristine atolls that make up the Bahamas, Caribbean islands are fragments of paradise cast adrift in a sea of problems.

The West Indies face difficult times. Agriculture, the economic backbone of many of its island nations, is under threat from the proposed withdrawal of preferential trade agreements. Agricultural land has also become a scarce commodity: farm plots are being subdivided into ever-smaller parcels, fallow periods are being progressively shortened, and marginal lands are being brought into production at a steadily increasing rate. Those who cannot find jobs in non-agricultural sectors continue to deforest the islands as they try to eke out a living from the land. Foreign aid, once an economic lifeline, is slowly drying up as the Caribbean is viewed less and less as a region in need of large-scale assistance. All of this is occurring at a time when the islands' populations and material expectations are steadily growing.

One organization has, however, taken a leading role in helping these island nations help themselves: RARE Center for Tropical Conservation.

RARE Center is a non-profit organization based in Arlington, Virginia. It develops innovative programs that protect endangered tropical wildlife and ecosystems. These programs are developed by RARE Center staff and a volunteer board of trustees made up of conservationists, scientists, and business people with broad experience in the tropics. RARE Center provides core funds to lead agencies in the communities from which a program is operated. Money to run these programs also comes from cooperating local businesses.

RARE Center believes that for conservation to become integrated into the day-to-day lives of people in less-developed countries, these people's natural resources must pay them tangible dividends: a good income and steady employment. At the same time people must understand how fragile their resources are, and how crucial it is to preserve those resources, in order to ensure that rural development is sustainable.

No matter how successful a country is in establishing reserves, changing legislation, or promoting a conservation ethic, these good works may be for nothing if population growth rates are not brought under control. A burgeoning population places enormous pressure on the ability of a country to both gainfully employ and satisfy the material expectations of its people. Unemployment precipitates environmental degradation as men and women struggle to make a decent living through forest clearance and subsistence agriculture.

Working with local people, RARE Center strives to change attitudes, promote economic linkages between environment and local economies, and work toward resolving the environmental problems caused by spiralling population growth.

## PROMOTING PROTECTION THROUGH PRIDE

RARE Center's Conservation Education Campaign (CEC) is a locally driven, one-year outreach program that uses marketing techniques, colourful flagship species, and national pride to generate grassroots support for conservation. CEC strives to build pride in and awareness of a target species – usually, an endangered bird – and to assist with the species's conservation. This work also provides a solid foundation for continuing outreach that promotes a more comprehensive environmental consciousness.

In twenty-five countries of the wider Caribbean, Central America, and the South Pacific, this program has reached out to more than 1.5 million people through fact sheets, posters, billboards, bumper stickers, costumes, songs, and school, community, and church visits.

In all but its Saint Vincent and Dominica pilot projects, local counterparts direct the program using core materials supplied by RARE Center. Its manual, *Promoting Protection Through Pride*, spells out different tasks that, when fully implemented, build upon one another to take the conservation message into every sector of the community. These tasks are designed to take place over the course of a single year. Questionnaire distribution and analysis allow workers to gauge existing levels of knowledge about the target species and its habitat and to monitor change in community awareness and attitude over the course of the project. The local counterpart implements the program in a specific area. These people do the day-to-day work of the program. They are usually employees of a local non-governmental organization (NGO), forestry department, or similar organization, which then serves as lead agency to the program. The lead agency oversees the implementation of the campaign and is responsible for logistical support, financial accounting, and providing a link between the counterpart and the local government.

If we see RARE Center's manual as a recipe book, and their supplies (badges, posters, etc.) as the recipes' ingredients, the lead agency is the "kitchen" in which the conservation "cake" is baked. The kitchen provides logistical and material support to the counterpart (the "chef"), covers the cost of vehicle maintenance, and pays the counterpart's salary. The lead agency also manages target species and the habitat in which it lives.

Local counterparts are encouraged to visit every primary and secondary school in their country or target area and to speak to as many children as possible. These talks introduce local children to the target species and to conservation issues. Puppets are often used to encourage younger children to participate in the campaign: the counterpart works with local teachers to develop their own puppet shows that, hopefully, will be used widely in the school system. Counterparts are also encouraged to solicit local assistance in producing a costume that depicts the target species and to produce a school song. Additionally, the production of an "A to Z" booklet of the host country's wildlife provides schools and schoolchildren with supplementary materials and resources that reinforce their interest in conservation.

Conservation-themed art or essay competitions are promoted to re-inforce and build upon the activities of a school visit. The donation of prizes by local businesses involves the larger community in the conservation campaign. RARE Center's manual recommends the production of a monthly or quarterly news sheet or comic book. These publications give children follow-up activities to complete, provide scope for corporate sponsorship, and may be used to continue outreach activities beyond the close of the project.

A CEC promotes its message to the wider community in a variety of ways. Talks and lectures to community groups, press releases, and articles and interviews for radio and television are prepared by the counterpart. As well, many visible reminders of the campaign are erected. For example, colourful posters are widely distributed in communities throughout the target area and placed in prominent sites such as supermarkets, bars, schools, health centres, and government buildings. These posters promote the conservation message and provide a visual reminder of the target species. Bumper stickers, distributed freely to vehicles throughout the target area, also promote the campaign message, and attract local corporate support thorough sponsorship. They are tangible evidence of community participation. Billboards placed at prominent road junctions are seen by a wide cross-section of the local community and provide an additional opportunity for corporate sponsorship. Stamps offer yet another colorful way of illustrating and promoting the target species and the conservation message. They can reach out not only across the host country, but to nations around the world as well. RARE Center's manual recommends the production of a series of stamps depicting the target species.

The manual also emphasizes the important role of the spiritual community and encourages counterparts to solicit the assistance of religious leaders, asking that, for example, leaders give sermons on environmental issues to their congregations. RARE Center also emphasizes the importance of ensuring that law enforcement officials are aware of existing environmental legislation. It recommends that counterparts produce booklets that summarize conservation laws and that these booklets be distributed to police officers throughout the target country or area.

The farming community is often at the forefront of environmental problems. Hosting meetings with pertinent farmers' groups and emphasizing the benefits of wise land husbandry, the mutual need for sustainable development, and the plight of the target species are recommended community outreach tasks. The airing of popular songs and music videos on radio and television by local musicians helps reach out to young people. This task attracts the nation's youth and carries the conservation message to them.

RARE Center has funded, or is currently funding, this type of conservation marketing project in St. Lucia, St. Vincent, Dominica, Montserrat, Grenada, Nevis, The Cayman Islands, The Bahamas, Bonaire, Jamaica, Anguilla, Trinidad, Tobago, Turks and Caicos Islands, Belize, Costa Rica, Honduras, Western Samoa, Pohnpei, Palau, Mexico, Yap, Fiji, and Kosrae.

## SECONDARY OBJECTIVES OF
## RARE CENTER'S CAMPAIGNS

RARE Center's campaigns are used not only to draw attention to the plight of a specific species but also to introduce the public to the habitat in which the species lives and to generate an appreciation for both the species and the habitat. The campaigns train local workers in a broad range of innovative outreach techniques that can be used to market other aspects of environmental protection. Depending upon the project site, the campaigns have other secondary objectives, including:

- Promoting the establishment of specific protected areas, national parks, or forest reserves that will benefit not only the target species but other plants and animals that share its habitat;
- Promoting a knowledge of, and appreciation for, existing reserves, national parks, and forest reserves;
- Building name recognition for the lead agency, in order to bring its work to public attention; and
- Building constituent support for initiatives such as the passage of pertinent legislation, the registration of captive wildlife, and other land use/wildlife regulations.

## IMPLEMENTING THE CONSERVATION EDUCATION CAMPAIGN

A critical first step, and a prerequisite in implementing RARE Center's CEC, is the selection of a target species. An interested local lead agency must also be chosen to oversee project implementation, and a capable counterpart must be appointed to carry out the CEC's various tasks.

Ideally, the target species should be endemic (it must symbolize the host country's or target area's uniqueness); reside in a critical habitat (it must provide a focus for the project); and be "marketable." The use of an existing national symbol has proven to be especially effective because it provides a strong link to nationalism and pride. In some countries, the existing national bird is well-suited to be RARE's target species. In Grenada, however, RARE Center proposed that the government change its national bird to RARE's target species. The government agreed to do this. In Anguilla, Palau, Yap, and Kosrae, local counterparts co-ordinated a nationwide campaign to elect a national or state bird, as a prerequisite to the campaign's commencement.

The CEC campaign strives to not only promote an awareness of the target species and its habitat but also to help the lead agency build name recognition for itself and prepare its officers for the continuation of parts of the program after RARE Center assistance ends.

The extent of day-to-day involvement by a lead agency will depend upon the counterpart. Where the counterpart is self-starting, motivated, and capable, for example, the lead agency's role will be minor. A less experienced counterpart, however, may require greater supervision and a more active role for the lead agency.

A counterpart must be highly motivated, outgoing, and willing to work long hours. RARE Center does not insist that counterparts be trained educators or wildlife specialists, although a basic appreciation for the environment and the problems facing it are crucial. While age and gender are not important considerations, good counterparts are usually young and single.

So far, counterpart accomplishments have been spectacular: reserves have been established in The Bahamas (Abaco), The Cayman Islands (Cayman Brac), Grenada, and Dominica. Wildlife legislation has been written or revised in St. Vincent, Montserrat, Kosrae, Yap, and St. Lucia. Endangered populations are rebounding in these areas.

### THE ESTABLISHMENT OF TOURIST TRAILS: A CONSERVATION TOOL

The RARE Center's program of trail financing is designed to contribute to the development of forest-based tourism. Carefully planned trails can generate revenue and jobs and preserve ecosystems that are critical to soil and water conservation and wildlife habitat.

While monitoring the development and construction of trails in Dominica and St. Lucia, RARE Center staff recognized the need for an easy-to-use trail design and construction manual. This manual assisted local forestry departments with the sighting, design, and building of trails and with subsequent interpretation and management. The manual is directed at mid-level technical officers, including Forestry or National Trust personnel, who are assigned to projects in trail design, construction, interpretation, or administration.

*Trails: Conservation that Makes Dollars and Sense*,[1] guides the reader, step-by-step, through the process of designing and constructing a park trail system. It aims to help resource managers maximize the economic benefits of tourism, while minimizing any negative environmental impact. It also strives to upgrade guide services and provide the necessary tools for the development of

Table 1    Revenues generated by tourist trails, and their impact on local economies

| Trail | Date Opened | Revenue Generated ($US) | Contribution to the Local Economy ($US) |
|---|---|---|---|
| Des Cartier (St. Lucia) | May 1996 | 240,000 | 1,320,000 |
| En Bas Saut (St. Lucia) | March 1998 | 138,040 | 1,173,000 |
| Little Water Cay (Turks and Caicos) | October 1996 | 71,928 | at least 1,000,000 |

quality, low-impact trails, thereby providing visitors with opportunities to see more of a country's natural wonders and bring in much-needed foreign exchange and jobs.

*Trails* was written and published, thanks to grants from the US Forest Service, the US Fish and Wildlife Service, Sven-Olof Lindblad's Special Expeditions, and The John D. and Catherine T. MacArthur Foundation. In August 1995, fifty copies of the manual were distributed freely throughout the English-speaking Caribbean. Manual recipients were encouraged to consider trail establishment and to use the manual's opening chapters to select suitable sites, do some surveying, and prepare funding proposals for trail construction.

Since the manual's distribution, RARE Center has assisted in the financing and construction of ten trails in the Caribbean and Micronesia. RARE's $500,000 investment in Caribbean nature trails has generated over $4 million for conservation and local communities in four years. In 2001 over ninety copies of the Spanish version of the manual were distributed throughout Latin America. As of May 2001, eleven trails had been built.

## THE PRESSING ISSUE OF
## POPULATION GROWTH

The effects of rapid population growth on the environment are particularly evident in small island nations where exploding populations are threatening to destroy ecosystems and undermine social and economic progress. RARE Center had an unusual opportunity to address this problem in the Eastern Caribbean nation of St. Lucia.

Over the past twenty years, RARE Center has worked with local organizations to develop model programs in conservation marketing and nature-trail development. These models are spreading throughout the Caribbean and, increasingly, around the world.

St. Lucia's population doubles every thirty to fourty years, and will likely exceed 200,000 by the year 2015. At this level of growth, massive increases are occurring in the working-age population. Current unemployment among the island's youth is high, and thus will only increase further. RARE Center thus pledged its assistance to the government of St. Lucia and the St. Lucia Planned Parenthood Association (SLPPA) to help deal with this growth through a Family Planning Initiative program. This assistance was made available to the community through a radio soap opera and an enhanced, field-based SLPPA outreach program. RARE Center believed that the soap opera would bring about a major shift in St. Lucian attitudes toward family planning. The SLPPA outreach program was designed to reinforce this change in attitude.

Because St. Lucia is a small island-based society, the power of a radio serial drama to change attitudes and reproductive behaviour is much clearer to demonstrate than in larger countries. RARE Center's Family Planning Initiative addresses a problem with grave social and environmental consequences and become an important case study.

When RARE Center's board of trustees embarked upon supporting the production of the soap opera *Apwe Plezi* (from the local Creole saying, "apwe Plezi c'est la pain" – after the pleasure comes the pain), it received encouragement from organizations such as the United Nations Family Planning Association (UNFPA), Johns Hopkins University, and Population Communications International (PCI). This latter organization generously provided advice from its researchers, scriptwriters, and production team. No easy-to-use, step-by-step manual, which would teach the team how to design and implement an issues-based radio drama was available to RARE Center or the St. Lucian partners, however. Thus, much on-the-job learning took place.

In early 1995, as a prerequisite to developing the drama, RARE Center reviewed existing literature and facilitated a series of twenty-eight island-wide focus-group meetings, as well as a questionnaire survey that sampled 1 per cent of the island's reproduc-

Table 2   Results of the RARE Center's survey assessing listener frequency of the radio serial drama *Apwe Plezi*.

The survey was designed to assess what listeners learned and how their attitudes changed on a wide range of issues related to family planning. Data below indicate the percentage of listeners who said that they had learned about a particular issue and/or that *Apwe Plezi* caused them to have a significant change in attitude about that issue.

| | Learning (per cent) | Change in attitudes (after episode no. 260) (per cent) | Change in attitudes (after episode no. 365) (per cent) |
| --- | --- | --- | --- |
| AIDS | 42 | 69 | 77 |
| Teenage pregnancy | 67 | 76 | 84 |
| Fidelity | N/A | 47 | 58 |
| Drugs | 60 | 60 | 74 |
| Spousal abuse | 52 | N/A | 61 |
| Rape | N/A | 56 | 65 |
| Contraception | 57 | N/A | N/A |
| Child spacing | 33 | 54 | 57 |

Responses do not add to 100 per cent because respondents could choose more than one factor. 13 per cent of respondents indicated that they learned nothing after listening to *Apwe Plezi*.

tive-aged population. The data collected gauged existing attitudes about family life, spousal abuse, and reproductive issues and helped craft the soap opera's message and story line. *Apwe Plezi* was first broadcast on 5 February 1996 amidst a local publicity blitz.

With its cast of good, bad, and ambivalent characters in recognizable scenes from everyday life, the drama showcased the rewards and successes of those who practise family planning and the negative consequences that befall those who do not. In its first year the program became the second-most popular program on the radio station, and the fourth most popular program in the country. Focus-group meetings, questionnaire surveys, satellite listeners, and telephone hotlines were used to evaluate the impact that this radio production had on the St. Lucian populace.

The second broadcast of *Apwe Plezi* began in July 1997 and drew to a close in September 1998 after 365 episodes. Once again a series of focus-group meetings and questionnaire surveys were held to gauge message penetration and subsequent changes in attitude, knowledge, and behaviour.

In the questionnaire survey administered after episode 365 was broadcast, 28 per cent of respondents considered themselves "infrequent" listeners of the program, tuning in less than once a week, while about 11 per cent were "regular" listeners, tuning in at least once per week. Half of the regular listeners were from rural areas and 34 per cent lived in urban areas. Female listeners outnumbered males by a margin of two to one and *Apwe Plezi* was their favourite program. Identification with *Apwe Plezi* characters was strong: 69 per cent of regular listeners said that they "sometimes felt like giving advice to the characters."

The survey designed to assess the educational impact of the show indicated that listeners learned about a variety of topics through the program, from AIDS to spousal abuse (see table 2). More importantly, perhaps, the respondents claimed that the serial brought about important changes in attitude, especially vis-à-vis AIDS, teenage pregnancy, and drug abuse.

When asked what they actually did as a result of listening to *Apwe Plezi*, 40 per cent of respondents said they spoke to a friend about issues raised in the program, 25 per cent said that they spoke to their partner, 13.5 per cent said that they had adopted some form of family planning, 4 per cent had arranged for family-planning counselling sessions, and 2 per cent said that they had visited the SLPPA. Of those who indicated that they had "talked to somebody else about *Apwe Plezi*," 48 per cent said they discussed a specific character's attitude or behaviour (up from 43 per cent in the last survey), 29 per cent talked about how a particular situation faced by the characters in the program related to their own life, and 20 per cent said that they discussed family-planning issues.

RARE Center believes that its programs need to be continually monitored and their results disseminated. The organization has therefore produced a series of publications that document the development and success of *Apwe Plezi*. This information can be found in its *Responsibility* reports, numbers one to eleven, which are available from RARE Center's Arlington office.

Other project components seek to improve the field-based outreach of SLPPA and reinforce gains made by the soap opera, including outfitting a vehicle with audio-visual equipment, developing effective visual materials such as portable billboards, and training SLPPA field staff in communication skills.

CONCLUSION

RARE Center believes that a holistic approach to conservation is crucial, since measures that merely address one part of a complex situation of environmental degradation will have little effect if remaining factors are not tackled simultaneously. Furthermore, for conservation to become a reality, action must come from within a region's community and programs must be implemented by local people who have knowledge, understanding, and concern for the ecological, social, political, and economic realities of their country. Working with local people and local organizations, RARE Center plays a small but vital part in helping communities help themselves.

NOTE

1   Paul Butler, *Trails: Conservation That Makes Dollars and Sense* (RARE Center, 1995).

## 9 Conclusion: Blending Universal and Local Ethics

### *Accountability towards Nature, Perfect Strangers, and Society*

ANIL K. GUPTA, VINEET RAI,
KIRIT K. PATEL, MURALI KRISHNA,
RIYA SINHA, DILEEP KORADIA,
CHIMAN PARMAR, PANNA PATEL,
AND HEMA PATEL

EDITORS' INTRODUCTION

In their conclusion to *Protecting Biological Diversity: Roles and Responsibilities*, Anil K. Gupta *et al.* look for solutions to ethical dilemmas inherent in our understanding of accountability – the notion that conservation biologists do not work in a vacuum but are accountable to nature and to society. The authors note that conservation biologists and ethnobiologists have not yet proved their willingness to conform to either local or universal values. They also remind us that the world's most biodiverse regions are inhabited by the poorest people. Their message, which pulls together the concerns of this book, is that the conservation of biological diversity should act to protect all beings, both human and non-human.

INTRODUCTION

Conserving our natural environment requires examining our perceptions of nature.[1] We don't seem to realize that our habit of responding emotionally to non-human, sentient beings mimics the rules of human social order. Animals and plants are judged using human understandings of what is good and bad, useful and non-useful, and desirable and undesirable. A good example of this tendency is the use of the term "weed," (a plant that is considered undesirable or out of place). Obviously, in nature no plant is out of place. Human beings either do not realize the significance of a plant in a particular place, or its appearance does not make sense. In some areas we have disturbed the environment so much that "undesirable" plants find it more convenient to grow than "desirable" plants, a fact that says nothing innate about the plants or their habitats, but does say something about the way in which we relate to our natural surroundings.

Another way to look at the metaphor of the weed is to ask, Can we ever locate a book in a library if the catalogue is lost? This situation is similar to the one that arises when we allow local knowledge about diversity to be lost through lack of recognition, respect, and failure to reward local experts.[2] In such cases, many plants in a forest become weeds simply because their catalogue or use has been lost.

When a cyclical view of life is taken, however, as it is, for example, in Hindu thought, one assumes that a human form is reached after one lives through 8.4 million lives or *yonis* (beings), ranging from ants, beetles, and micro-organisms to large mammals. Responsibility toward nature stems from one's vision of sharing a common life cycle with other species. One's own identity, then, as expressed through language, culture, and social institutions, is not so far removed from the identity of other living beings. In fact, a *vedic* hymn requires prayer to be performed for the well-being of all living beings, not just just the followers of one's sect or other humans. This kind of practice is one way to understand and respect universal ethic. There are also those (ecologists, for example), who believe that nature possesses its own logic, which we can understand only partially and which we cannot replace with human social logic. Individuals' ethical positions vis-à-vis nature vary within this range.

The current thinking in conservation biology suggests that we should preserve biodiversity for future generations (see chapter 6). The future generation is made up of perfect strangers, however, and despite numerous tales and fables in which animals talk not only amongst themselves, but also to us, our understanding of wildlife is much like our understanding of strangers – they are unknown and unknowable. Our responsibility toward those not yet born or towards beings whom we do not understand as capable of having feelings cannot, therefore, arise entirely from utilitarian logic. This concluding chapter seeks to establish the grounds on which we can build responsibility toward nature.

Ecological economists use the terms option value and existence value to talk about non-utilitarian value of objects and biodiversity, and potential value to indicate the potential returns from a resource, as a function of the returns obtained in the past. For example, the option value of a tropical forest includes the probability of a drug being found in its constituent plants. If we found five drugs in 5,000 plants screened from a forest, we could assume that we might discover drugs in other forests. The value of the drugs might then be attributed as the option value of other similar forests. Existence value refers to the intrinsic value of a resource, not a present or future utilitarian value. For example, the Taj Mahal or the panda are valuable because they are rare and unique.

To assess the utilitarian value of biodiversity, ecological economists use the terms exchange value and use value. Exchange value is the value that is attached to the exchange of natural resources extracted from a given region. The use value is the benefit that is derived by various users from exploited resources.

Obviously, any resource may be valued under the afore-mentioned system, and an accounting of our responsibility may accordingly take place. The public sector and the private sector may also take different values into account when allocating resources towards conservation, and society must represent multiple concerns. The challenge for biodiversity conservation today is the generation of a coalition of interests that will enlarge the space in civic consciousness for conservation, particularly of those components of which we know little or nothing.

A potential problem with the attribution of monetary value to biodiversity stems from the argument that if something of value is obtained from a natural habitat, human tendencies are such that the resource is likely be overexploited.[3] The logical implication of this argument, then, is that, in order to conserve biodiversity people should be kept poor, a position that is neither ethically nor sociopolitically acceptable. Instead, institutional choices that help improve local community's prospects by generating opportunities to derive higher economic benefits from their environment's conservation should be promoted. Which choices are generated by different stakeholders will inevitably depend upon our ethical accountability towards nature, society and the next generation.

In a recent paper, Anil K. Gupta identified seven kinds of ethical accountability that are relevant to this discussion:[4]

1 The accountability of both public- and private-sector researchers and biodiversity prospectors working in national or international organizations toward providers of biodiversity resources from wild, domesticated, and public access domains;
2 The accountability of researchers and prospectors toward the host country;
3 The accountability of professionals toward academic communities and professional bodies guiding the process of exploring or extracting biodiversity;
4 The accountability of international, United Nations, or other organizations that possess globally pooled germ-plasm collections deposited in good faith but accessible to public or private institutions without reciprocal responsibilities;
5 The accountability of institutions of governance that legitimize various kinds of property-right regimes that lead to various ethical and moral dilemmas;
6 The accountability of society and consumers of products derived from prospected biodiversity or competing alternatives; and
7 The accountability of conservationists, users, and consumers toward future generations and living, non-human, sentient beings.

Two other kinds of accountability also seem relevant to this discussion:

8 An accountability toward nature, including plants, animals, and other forms of life, as well as their habitats; and
9 An accountability toward our consciences, as well as toward universal ethical values.[5]

## THE INTERNATIONAL PERSPECTIVE IN ACCESSING BIODIVERSITY: NORTH-SOUTH RELATIONS

Research collaborations between local communities and outside researchers often involve a dilemma that has already been brought into sharp focus (Project Camelot is a good example.)[6] Important issues of covert and overt research, inadequate provision of information to respondents, information obtained through deceit, and violation of local cultural and spiritual beliefs during the acquisition of information or material have all been noted. In this book, not all of these dimensions have figured equally prominently, but many of them are apparent. In the following section, we highlight some key issues emerging from various chapters of this book and present a way in which we think the dilemma emerging from the discourse could be resolved. The discussion has been organized around three kinds of accountability.

Three things must be kept in mind when considering researchers' accountability: the responsibility of national and international researchers toward local communities differs only in degree, not quality; the poor are not better off being exploited by national researchers or institutions than by international institutions; and responsibility for conservation is higher, for national researchers of both private and public institutions, than their international counterparts.

### Accountability of Researchers Toward Information Providers, their Communities, and their Countries

Values determine the choice of goals toward which an effort for change is directed. Any intervention is designed to maximize a particular set of values while minimizing the cost of other values.[7] Conservation researchers practice social intervention through conservation projects, with the goal of conserving biodiversity. It is important that they simultaneously try to minimize losses to local communities, as communities are sometimes uprooted from their traditional land as a result of policies suggested by conservationists. Such uprooting may destroy a community's knowledge about the local biodiversity, knowledge that is priceless.

This book provides many examples of unethical researcher behaviour, and several authors suggest ways in which to improve relations between researchers and local people. Listed below are some of the most problematic areas in this book, as well as some culled from our own experience and knowledge.

- Researchers have both inadvertently and knowingly passed judgment over the values and culture of local communities in their host countries. They have also, at times, violated norms of social intimacy, causing permanent strain in the researcher/community relationship (see chapter 5).

  Respect for local culture, customs, and institutions may also require respect for local language. Many, if not most, researchers never discuss their research agenda or findings with local communities in the local language. Dr. Rakotovao *et al.* (chapter 5) have rightly emphasized the importance of using local dialect in all communications.

- Foreign researchers should be obliged to share a copy of their research material (field notes, photographs, diskettes, samples, and other information) with local institutions and counterparts. To be most useful to local communities, the publication policy must evolve so that the interests of local communities are not overridden by the interests of scientists.[8]

- The sharing of credit for intellectual contribution among expatriate and local researchers has been a serious problem in most Southern countries: the colonial mode of extraction continues in most of the academic world. There are, however, exceptions to this norm: some researchers decide not to be first author of any publication when in another country, for example, so that their institution-building role is not compromised.

  Through their actions, researchers may legitimize decision-making processes in which local communities are not included. In most Southern countries, the elite, bureaucracy, and technocracy often see statements from international scientists and institutions, such as the World Bank, as a vindication of their policies and perceptions. Ethical conflicts are inherent when local communities and non-governmental organizations (NGOs) oppose the very same policies and projects. There is no simple solution to this problem, so expatriate researchers must consider the extent to which their actions and decisions affect the lives of local people.

- In some cases, researchers must provide socio-political and economic support to local communities (see chapter 6). This may cause tension in the sensitive relationship between expatriate researchers and local peoples. In an oppressive regime, support to local communities invites the charge of insurgency, yet participation and legitimization of such a regime may be equally unethical. The conflicts between local communities and national governments are better resolved within national boundaries, and foreign researchers should never be involved in such conflicts. Where help ends and hegemony begins is very difficult to determine if involvement of foreign researchers becomes acceptable and desirable.

- National researchers working at local institutions are, typically, obliged to share the intellectual property of their findings with the state or with other supporters of national institutions, a requirement that can generate disagreeable consequences. For example, several researchers in India felt unhappy about contracts between national institutions and foreign agencies that required provision of access to local biodiversity in lieu of aid.

- In chapter 8, Paul Butler and Victor Alleyne Regis demonstrated the importance of commercializing local resources as a way of generating spillover effects and benefit sharing. The idea of promoting protection through pride is extremely positive and worth careful consideration. Conservation education should become part of every conservation biology project so that reciprocity in learning is ensured and institutionalized.

- In chapter 1, conflict over the use of project resources by expatriate scientists vis-à-vis local communities is highlighted in the example of Virunga National Park. Such conflicts are not specific to natural resource-based projects, of course – they can arise in any project situation. Léonard Mubalama looked at the wholesale violation of the terms of a research agreement by foreign scientists in the République démocratique du Congo. The scientists removed all of the project samples from the country without authorization, the local researchers and communities obviously felt betrayed when their rights were trampled upon in this way.

The responsibility of researchers to present their work to local communities in an easily comprehensible manner is not merely ethical: it is also institutional. How else can local communities enlarge their understanding of and, consequently, their responsibility toward, nature? This question is highlighted by Bonarge Pacheco, *cacique* (chief) of the Emberá people of Ipetí, Panama, in an interview with Rogelio Cansarí, (see chapter 3).

*Responsibility of Researchers and International*
*Organizations Toward their Profession*

In conservation biology and ethnobiology, standards of account-ability towards one's peers have not yet been clearly outlined. Some professionals have developed codes of conduct, but their mecha-nisms for enforcing the codes are often very weak. For example, a researcher can present a paper in a conservation biology conference without sharing the findings with local communities. Similarly, a national or corporate gene bank in a Northern country may accept an accession from a scientist without confirming whether the mate-rial was obtained legally and in a morally acceptable manner. Patent offices can issue patents to scientists without ensuring that the patentees declare lawful and rightful property rights over the invention.[9]

Standards of good practice have been defined in several profes-sions, but professionals frequently forget that these standards should also be applied when dealing with non-professionals. For example, it is accepted practice that any communication that has substantive implications for ideas should be acknowledged. Accordingly, personal communications find place in academic dis-course. This accountability is generally observed only towards pro-fessional colleagues, however. It is extremely rare that farmers, Indigenous people, artisans, or others who have working knowl-edge of certain problems are ever acknowledged. In fact, the entire discipline of ethnobiology has gained legitimacy through extraction of information without proper acknowledgement. As well, the wealth accumulated from this knowledge is seldom shared with the providers.

In chapter 2, Marie-Hélène Parizeau raises some very pertinent questions. The exclusion of local communities from protected areas has often been advocated by conservation biologists as an example of an ecologically, socio-economically, and culturally hurtful policy. Similarly, when turtles are released from fishers' nets without their knowledge, what was the right course of action to be taken? As well, sharing the profits of a pharmaceutical company's new drug development with all the countries and communities that provided research opportunity for identifying the raw material may be a step in the right direction,[10] although whether this scale of sharing is suf-ficient or not is debatable.

These kinds of ethical difficulties prompt Dr. Parizeau to argue for the evolution of a professional code of conduct for conservation biologists. Her valid contention is that, without guidelines, easily avoidable mistakes are being made. Thus, the evolution of professional guidelines is a useful goal and must be pursued collectively. A proposed guideline by the American Society of Economic Botany states that a researcher should do everything in his or her power to ensure that local communities that provide knowledge and resources receive their due share in proceeds from the commercialization of biodiversity. The equitable sharing of benefits is also enshrined in the Convention on Biological Diversity (CBD) as one of its three fundamental goals. Every professional must be aware of the responsibility.

In chapter 3, Rogelio Cansarí gives a very interesting account of the efforts of an Indigenous community in Panama to ensure that outside researchers behave ethically. In Emberá regions, after long struggles, local institutions declared all land to be common property. The countless negative experiences of Emberá communities with foreign researchers led to a serious rethinking of the relationship between the local community and the scientists, as the communities could neither question the methods, the purpose, nor the process of enquiry followed by outsiders. The Emberá hosted and assisted many scientists in the field, and many promises of future reciprocity were made but never kept. The consequence of such irresponsibility is, often, the development of a strict regulatory regime that may hurt the work of those scientists working toward the well-being of local communities. A permit is now required to visit and meet local Emberá communities as well as the regions where no Emberá live.

The fact that Emberá communities did not learn anything of value from visiting scientists is distressing. The least that any visiting scientist should do is to share his or her knowledge with local communities. Law #22, passed by the Emberá Congress in 1983, has had a significant impact on the relationship between researchers and Emberá communities by forcing a measure of respect for local community leaders. The steps taken illustrate how the relationship between local communities and expatriate researchers is likely to be shaped. Obviously, a profession that does not generate goodwill and respect is bound to eventually suffer from the same kinds of conflict it has imposed on others. Conservation biologists, ethnobiologists and anthropologists have not as yet proved their willingness to modify their way of working, and to conform to local and universal ethical values.

In chapter 7, Georgina Wigley and Heather Baser examine issues that arise in the context of international development. They examine Organization for Economic Co-operation and Development (OECD) and other guidelines, and highlight the ethical responsibilities of researchers in an international context. A model profile of a technical adviser is outlined by the author. They emphasize the role of interpersonal and communication skills, as well as competence in technical areas, but do not distinguish the ethical perspective from the institution-building perspective. For example, when a well-paid researcher or expert lives and works with less-entitled counterparts, some distance is bound to exist between the parties. If the technical adviser spent considerable time building the capacity of local institutions and individuals without taking undue credit, however, status differences might dissolve.

## Accountability Toward Our Consciences

In chapter 4, Priscilla Weeks, Jane Packard, and Mirella Martinez-Velarde illuminate the way in which negotiations among researchers and local communities in host countries can be pursued in a way that generates mutual respect and professional co-operation. Many conflicts that arise between local communities and foreign researchers are similar to the issues that arise out of internal feelings of professional responsibility. The authors critique the conduct of both local and international professional communities, and recognize the inherent risk of miscommunication and mutual disrespect in collaborative research among local and expatriate researchers when the motives and rewards of each party are not properly articulated or respected. The attitudes of "hit-and-run" researchers who are so concerned with their own professional growth that they do not feel accountable to local communities generate tremendous misgivings for both local communities and researchers.

At the same time, the authors highlight remedies that can typically be carried out without loss of academic rigour; such as (a) the contribution of time and resources by expatriate researchers for capacity building in the host country; (b) the reordering of research priorities to suit local needs; (c) the need for international institutions to explain internal conflicts and constraints that impede the fulfillment of their responsibilities; (d) the recognition that technical priorities can be fulfilled within cultural and institutional contexts; (e) the overcoming of the "tool bias" (the assumption that one's own tools or technical or academic profession are the most important, relevant, and desirable tools with which to solve a problem); and (f) the sharing of information that might serve the purpose of conservation as well as the profession, such as the precise location of endangered wildlife.

Researchers often are dismayed by the way in which their work has been interpreted when they have no control over the use of their data and conclusions.[11] Similar concerns have not been expressed about how Indigenous communities might think about the same problem, however. W.E. Hopkins suggests that "in normative terms, when any two individual cultures have differences regarding the morality of a particular action or behavior, both can be right because morality is relative. This sense of moral relativism suggests that absolute notions of right or wrong are not valid."[12] He implies that the notion of universal morality is invalid, and that a universal value system does not exist strictly within certain specified limits. Differences in cultural diversity should not be used as part of an argument for total relativism in moral values. To take something such as biodiversity or related knowledge from someone who is not aware of its true worth without due consideration and informed consent might be considered fraud, something no society can condone as a legitimate or fair activity.

### THE ETHICAL CONTEXT OF THE CONVENTION ON BIOLOGICAL DIVERSITY

The Pew Conservation Scholars developed ethical guidelines based on Anil K. Gupta's first three dimensions of ethical responsibility. Their Suggested Ethical Guidelines recognize that local communities, as well as researchers and corporations, have used prospected biodiversity (that which is used to develop commercial products or services) for a long time and that the conservation of cultural and biological diversity is closely intertwined.[13] The guiding principles recognize that (a) research is an educational process for all involved (even if opportunities of learning may not always be reciprocal or balanced); (b) proprietary rights over scientific knowledge are not fundamentally different from the rights of producers and providers of traditional knowledge and contemporary innovations; (c) local cultural values and norms must be respected and fair and equitable sharing of benefits among various stakeholders must be achieved.

The Guidelines deal with four kinds of relationship between researchers and the local communities: (a) non-extractive and non-commercial; (b) extractive but primarily non-commercial; (c) non-extractive possibly commercial; and (d) extractive for commercial purposes. Ethical obligations are different in each relationship and, consequently, certain aspects of the Guidelines are phrased using the words "must," "should," and "may." The scholars realize that different professionals and political communities may have a genuine difference of opinion on these guidelines but hope that they can provide ground for further progress.

Regions of high biodiversity should not remain regions of high poverty, illiteracy, or unemployment, and labour-participation rates for women and children should increase. This nexus can only be broken if an appropriate arrangement for sharing benefits derived from biodiversity exploitation is put in place. If the current extraction practices keep people poor, how much more biodiversity will disappear before corrective actions are taken? The CBD provides a framework through which the current imbalance in responsibility can be corrected.

Article 15.5 of CBD, for example, requires the informed consent of host governments prior to obtaining access to genetic resources. The Pew Scholars and other conservation guidelines require prior approval from the appropriate national authorities, including Indigenous people or local communities. The consent of communities or innovative individuals can be invoked only through Article 8(j).[14]

Putting these articles into practice will require several important provisions. In the wake of the Merck-InBio deal,* the expectations of states and communities have increased dramatically. Moderating these expectations in some cases may become imperative. On the other hand, many communities do not even know the value of their

---

* This deal, signed between Merck and the National Institute of Biodiversity (InBio) in Costa Rica, helped InBio build its capacity to make an inventory of local biodiversity through Merck's collection of biodiversity samples from unpopulated lands.

knowledge or their resources, and may prefer not to assign material values to their ethereal relationships with nature. Other challenges include determining representative structures of people when administrative boundaries do not overlap with the ecological boundaries of a resource, and determining the time frame over which benefits should be shared.

How much information is sufficient and when negotiations among people and outsiders should be considered satisfactorily concluded will become clear only through experimentation. Clearly, researchers cannot take advantage of the generosity of local communities and individuals. The Pew Scholars' Ethical Guidelines clearly distinguish four stages in the negotiation of the terms of access to local biological resources: (a) when access occurs; (b) when a new use is discovered; (c) when a product is developed; and (d) when commercialization occurs. Of course, each party also needs an assurance that negotiations will move in a stage-by-stage manner. Several safeguards have been suggested, and these can help moderate mutual expectations and generate reasonable rewards:

1 An international registry of innovations, both in the name of individuals and communities, has been suggested.[15] Registration should assign a right of precedence and protection for a limited period, during which either other communities or institutions could claim to have independently developed the innovation or would agree to share derived benefits. An international fund under CBD or CPGR (Commission on Plant Genetic Resources, FAO (Food and Agriculture Organization)), could help maintain the registry in collaboration with the World Intellectual Property Organization. The system should allow communities to set up collective funds for local conservation and economic well-being. SRISTI (Society for Research and Initiatives for Sustainable Technologies and Institutions) has argued that the rights of individual innovators or conservators of resources should be specifically protected even if they do not pursue the same in the short run. Whenever a reward becomes due, the innovator or conservator concerned should have the option of deciding what to do with the material resources when they become available.
2 Individual innovations should not be subsumed under traditional knowledge. This tactic is often attempted by NGOs and international organizations. The HoneyBee Network, supported by

SRISTI, has thousands of innovations in its database that can be traced back to their specific innovators, for example. Many of these innovators did not develop their material for financial gain, but this is no reason for denying them due reward. Similarly, individual contributions to conservation should be distinguished from collective contributions. Only a small section of a village may be growing land races, for example, and so making the entire community custodian of any reward would be unfair. If the majority of the people (i.e., more than 75 per cent) grow land races, however, the community-level reward may make sense.

3  No conservation-incentives scheme should lead to the erosion of the natural resource base for which the incentives were put in place. Some people argue that providing material incentives may distort the values of the local communities, who are supposed live in harmony and peace with nature. Material rewards in the absence of local institution building can indeed lead to environmental and cultural degradation: in many North American Indian reservations, for example, the social welfare system, unsupported by investment in local institution building, has killed the spirit of local enterprise. There are communities such as the Zunis, however, who have won major lawsuits and have obtained substantial monetary compensation to undo the damage to their natural resources that resulted from unauthorized dumping by the state. These communities are using recent technology (such as GIS, the Geographical Information System) to manage natural resources optimally and are reviving some old technologies and land-use systems in order to rejuvenate the irrigated lands.

The absence of monetary reward and other opportunities is unlikely to either preserve the resource or the ethics that have helped to conserve the resource so far. The following matrix (see Table 1) helps combine material and non-material incentives to conserving biodiversity, rewarding creativity and innovation, generating respect for local institutions and ethical behaviour, and influencing the values of future leaders.[16]

The first category, individual material rewards, includes conventional incentives such as patents, licence fees, contract fees, and monetary rewards for innovations and conservation efforts. The innovators must decide what to do with their reward. For example,

Matrix for Combining Material and Non-material Incentives to Conserve Biodiversity

|  | Form of Rewards | |
|---|---|---|
|  | *Material* | *Non-material* |
| **Target of Rewards** | | |
| Individual | 1 | 2 |
| Collective | 3 | 4 |

some individual innovators have refused any private reward. In such cases, one can set up a trust fund for collective use of the reward money, under the leadership of individuals whose contributions made this possible. Such a measure generates non-material individual reward in the form of honour or esteem. The consumers of biodiversity-based products also have to shoulder some responsibility for its erosion. By not ensuring that the companies that market biodiversity-based products share benefits with the local communities that conserve biodiversity and related knowledge, consumers contribute to the exploitation of these communities and their knowledge systems. Just as animal rights activists ensured that manufacturers of cosmetics no longer test their products on animals, consumer awareness can create pressure on the biodiversity-based product manufacturers and marketers to become more responsible. Incentives in the form of training, better prices, better ways to meet basic needs, and education can be provided by those who make use of biodiversity for commercial reasons.

The second category, non-material individual rewards, includes honour, recognition, and respect for individuals who have contributed extraordinarily to the goals of conservation or value addition. SRISTI has honoured about fifty such individuals so far in the state of Gujarat in India. They have also organized biodiversity contests among schoolchildren and honoured the most knowledgeable children. Small prizes and certificates contribute to building respect for local knowledge. Conservation through competition has been a very successful experiment, and has been pursued by SRISTI in different parts of India and around the world.

The third category, collective material rewards, offers enormous scope for innovation. Several kinds of trust funds and guaranteed risk, or venture capital funds, can be set up to promote conservation, value addition, commercialization, and other goals. These funds should provide enough flexibility for communities to pursue culture-specific norms of conservation as well as offer reward and/or compensation to outstanding local contributors. Some of these funds will operate regionally, while others may be implemented at the community level.

Finally, the fourth category, non-material collective rewards, includes policy reform, institution-building, incorporation of local ecological knowledge in the educational curriculum at different levels, and development of markets for organic and other local products at both the national and global levels. Although no one incentive may be enough to generate the right kind of respect for traditional knowledge and contemporary conservatory innovations, a combination of these incentives can provide positive, sustainable outcome.

### SUMMING UP

A review of ethical dilemmas and value conflicts revealed many areas in which external scientists could develop more transparent criteria of their effectiveness in dealing with the sources of these conflicts.[17] The number of personal communications cited in Northern publications was not found to be substantially different than in some Southern publications. A feeling remains, however, that Southern scholars may not be cited in the Northern work as often as Northern scholars, other things being equal. Obviously, such an impression cannot be removed easily. Similarly correcting this bias will not automatically resolve the problem of Southern professionals who do not cite compatriots or local communities even when they learn unique insights from them. Greater accountability from within comunities is thus also an important goal.

The collection of essays in this book mark an important landmark in the discussion of conservation biologists' professional accountability toward researchers, local communities, professional bodies, nation-states, nature, and future generations, both in Northern and Southern countries. It provides starting points for a much-needed discussion of the ethical dilemmas that arise within Western science and scholarship on historical conduct and current professional values. For example, the tradition of guarding an informant's identity in order to protect him or her from any negative reaction from colleagues, fellow citizens, or leaders, may need to be re-written in the post-CBD context. In one of SRISTI's projects, supported by IDRC (International Development and Research Centre), such an issue emerged because the IDRC guidelines required a similar protective principle. SRISTI had to assert that its goals required every individual contribution made by the local communities and individuals to be cited unless otherwise requested by the respondent. This suggestion was accepted.

The competition between historical and current preferences, as discussed by Bryan G. Norton in chapter 6, must be accepted as a source of necessary tension in our daily professional pursuits. The diversity in nature need not necessarily generate diversity in all the values of conserving nature although, of course, diverse ways of resolving conflicts and dilemmas must remain.

The least explored in this book is that of our responsibility toward future generations, although Bryan G. Norton touches on it in chapter 6. These strangers' voices aren't yet heard, and so their needs and preferences must be inferred, anticipated, and responded to by the present generation, using contemporary as well as traditional value systems. Inequity in the present generation is thus superimposed upon and carried forward into the next generation.

Human needs cannot always take priority over the needs of nature and other living beings. Problems arise when those who gain the most (i.e., tourists, scientists, or consumers of herbal drugs or products) from an area's biological diversity, are also those who contribute the least to the alleviation of suffering and poverty of the communities living in and around these areas. Larry Merculif perhaps states this problem most clearly

They [animal right activists] do not understand, in their desire to protect animals, they are destroying the cultural, economic, and spiritual systems which have allowed humans and wildlife to be sustained over thousands of years . . . Their [Animals First activists] concept is based upon a belief that animals and humans are separate and they project human values onto animals. Ours is based on knowledge from hundreds of generations, which allows us to understand that humans are part of all living things and that all living things are part of us. As such it is spiritually possible to touch the animals' spirits in order to understand them. Our relationship with animals is incorporated into cultural system, language and daily lifestyles. Theirs is based on laws and human compassion . . . Because we are intricately tied to all living things, when our relationship with any part of such life is severed by force, our spiritual, economic, and cultural systems are destroyed. Deep knowledge about wildlife is destroyed; knowledge which western science will never replace . . . I leave you with this last thought: We have an obligation to teach the world what we know about proper relationships between humans and other living things.[18]

The continuity between human and non-human life is a new discovery for contemporary cultures, but has been part of everyday experience for many Indigenous communities for countless generations. Ethical dilemmas are like the Plimsoll line of a ship: unless one deviates too much from this line, the ship does not sink. But if we wait, without intervening, the ship will eventually sink.

### NOTES

1  A. Gupta, "Ethics of Foreign Aid: Why Is It Always Ignored?" in *Criteria for Foreign Aid* (Aarhus, Denmark: University of Aarhus, 1992).
2  Society for Research and Initiatives for Sustainable Technologies and Institutions and HoneyBee Network, *Annual Reports Of Society for Research and Initiatives for Sustainable Technologies and Institutions*, 1993–1999, (Ahmedabad, India: SRISTI, 1999). See also http://www.sristi.org.
3  J. McNeely, "Economic Incentive for Conserving Biodiversity: Lessons for Africa," *Ambio* 22, nos. 2–3 (1993): 144–50.
4  A. Gupta, *Compensating Local Communities for Conserving Biodiversity: Shall We Save the Goose that Laid the Golden Eggs So Long?*, IIMA (Indian Institute of Management, Ahmedabad) working paper no. 1206, Ahmedabad, India, August 1994.
5  HoneyBee Network, *A Newsletter Dedicated to Networking Grassroots Innovators and Other Knowledge Rich Economically Poor Creative Communities and Individuals Around the World* (Ahmedabad, India: HoneyBee, 1999); and SRISTI, and HoneyBee Network. See also http://www.sristi.org.
6  I.L. Horowitz, *Rise and Fall of Project Camelot; Studies in the Relationship Between Social Science and Practical Politics* (Cambridge, MA: MIT, 1974), 409.
7  H.C. Kelman and D.P. Warwick, "The Ethics of Social Intervention: Goals, Means, and Consequences," in Bermant, Kelman, and Warwick, eds., *The Ethics of Social Intervention* (Washington, DC: Hempshire Publishing Corporation, 1978) 4.

8 Gupta; and RAFI (Rural Advancement Foundation) Communique Endod,
   *A Case Study of the Use of African Indigenous Knowledge to Address
   Global Health and Environmental Problems*, 1–4 March 1993.

9 RAFI Communique, *Patents, Indigenous People, and Human Genetic
   Diversity.* May 1–6, 1993.

10 C. King, "Biological Diversity, Indigenous Knowledge, Drug Discovery
   and Intellectual Property Rights: Creating Reciprocity and Maintaining
   Relationships, *Intellectual Property Rights and Indigenous Knowledge*,
   October (1993): 1–27.

11 R. Moore, "Becoming a Sociologist in Sparkbrook;" and L.D. Cain,
   "The AMA and the Gerontologists: Uses and Abuses of "A Profile of
   Aging: USA," in G.C. Wenger, ed., *The Research Relationship* (London:
   Allen & Unwin, 1969), 59.

12 W.E. Hopkins, Ethical Dimensions of Diversity (Thousand Oaks, CA:
   Sage Publications, 1997), 84.

13 J.A. McNeely, "Common Property Resource Management or Government Ownership: Improving the Conservation of Biological Resources,"
   *International Relations* 10, no. 3 (1991): 211–25.

14 Article 8(j) of the Convention on Biological Diversity states: "Each
   Contracting Party shall, as far as possible and as appropriate, subject to
   its national legislation, respect, preserve, and maintain knowledge, innovations and practices of indigenous and local communities embodying
   traditional lifestyles relevant for the conservation and sustainable use of
   biological diversity and promote their wider application with the
   approval and involvement of the holders of such knowledge, innovations
   and practices and encourage the equitable sharing of the benefits arising
   from the utilization of such knowledge, innovations and practices."

15 A. Gupta, *Ethical issues in Prospecting Biodiversity*, IIMA, working paper
   no. 1205, Ahmedabad, India, (1994); and Society for Research and Initiatives for Sustainable Technologies and Institutions and HoneyBee
   Network, 1993–1999.

16 A Gupta, "Why Does Poverty Persist in Region of High Biodiversity?:
   A Case for Indigenous Property Rights System," paper presented at the
   International Conference on Property Rights and Genetic Resources,
   sponsored by IUCN (International Union for Nature Conservation),
   UNEP (United Nations Environment Program), and ACTS (African Centre
   for Technology Studies) (Kenya: 10–16 June 1991); A. Gupta, "Dilemma
   in Conservation of Biodiversity; Ethical, Equity and Moral Issues –
   A Review," paper prepared for a workshop of Pew Conservation
   Scholars on Developing Ethical Guidelines for Accessing Biodiversity
   (Arizona, 1994), published as "Ethical Dilemmas in Conservation of
   Biodiversity: Toward Developing Globally Acceptable Ethical Guide-
   lines," *Eubios Journal of Asian and International Bioethics* 5 (1995):

38–40; and A. Gupta, K. Patel et al., "Participatory Research: Will the Koel Hatch the Crow's Eggs?," paper presented at the International Seminar on Participatory Research and Gender Analysis for Technological Development, organized by CIAT (Colombia, 1997).

17 A. Gupta, "Ethical Dilemma and Value Conflicts," in *Management Research: A Review* (Ahmedabad, India: IIMA 1986).

18 From A. Gupta, *Sustainability through Biodiversity: Designing Crucibile of Culture, Creativity and Conscience.* IIMA working paper no. 1005, presented at International Conference on Biodiversity and Conservation, Danish Parliament (Copenhagen, 8 November 1991).

# Contributors

CHANTALE ANDREANARIVE  Centre national de recherche en environnement, Antananarivo, Madagascar

HEATHER BASER  Agence canadienne de développement international (ACDI), Hull, Quebec

PAUL BUTLER  RARE Centre for Tropical Conservation, Castries, Saint Lucia

ROGELIO CANSARí  McGill University, Montreal, Quebec

ANIL K. GUPTA  Centre for Management in Agriculture, Indian Institute of Management, SRISTI, and the Honeybee Network, Ahmedabad, India

DILEEP KORADIA  Centre for Management in Agriculture, Indian Institute of Management, SRISTI, and the Honeybee Network, Ahmedabad, India

MARGARET KRAENZEL, McGill University, Montreal, Quebec

MURALI KRISHNA  Centre for Management in Agriculture, Indian Institute of Management, SRISTI, and the Honeybee Network, Ahmedabad, India

MIRELLA MARTINEZ-VELARDE   Universidad Santa Maria la Antigua, Panama

LEONARD MUBALAMA   Institute Zaïrois pour la conservation de la nature, Ituru, République démocratique du Congo

BRYAN G. NORTON   School of Public Policy, Georgia Institute of Technology, Atlanta, Georgia

JANE PACKARD   Texas A&M University, Houston, Texas

MARIE-HÉLÈNE PARIZEAU   Université Laval, Quebec, Quebec

CHIMAN PARMAR   Centre for Management in Agriculture, Indian Institute of Management, SRISTI, and the Honeybee Network, Ahmedabad, India

KIRIT K. PATEL   Centre for Management in Agriculture, Indian Institute of Management, SRISTI, and the Honeybee Network, Ahmedabad, India

HEMA PATEL   Centre for Management in Agriculture, Indian Institute of Management, SRISTI, and the Honeybee Network, Ahmedabad, India

PANNA PATEL   Centre for Management in Agriculture, Indian Institute of Management, SRISTI, and the Honeybee Network, Ahmedabad, India

CATHERINE POTVIN   McGill University, Montreal, Quebec

VINET RAI   Centre for Management in Agriculture, Indian Institute of Management, SRISTI, and the Honeybee Network, Ahmedabad, India

R. RAKOTOARISEHENO   Centre national de recherche en environnement, Antananarivo, Madagascar

LALA H. RAKOTOVAO   Centre national de recherche en environnement, Antananarivo, Madagascar

VICTOR ALLEYNE REGIS   RARE Centre for Tropical Conservation, Castries, Saint Lucia

GILLES SEUTIN   Parks Canada, Hull, Quebec

RIYA SINHA   Centre for Management in Agriculture, Indian Institute of Management, SRISTI, and the Honeybee Network, Ahmedabad, India

PRISCILLA WEEKS   The Environmental Institute of Houston, Houston, Texas

GEORGINA WIGLEY   Agence canadienne de développement international (ACDI), Hull, Quebec